The Complete Guide To
Growing Berries and Grapes

ALSO BY LOUISE RIOTTE

In Nature's Hands
Carrots Love Tomatoes
Roses Love Garlic
Sleeping with a Sunflower
Astrological Gardening

The Complete Guide to

GROWING

BERRIES

and GRAPES

LOUISE RIOTTE

with Illustrations by the Author

TAYLOR PUBLISHING COMPANY
DALLAS, TEXAS

Copyright ©1974, 1993 by Louise Riotte
All rights reserved.

No part of this book my be reproduced in any form without written permission
from the publisher.

Published by Taylor Publishing Company
        1550 West Mockingbird Lane
        Dallas, Texas 75235

Designed by David Timmons

Library of Congress Cataloging-in-Publication Data
Riotte, Louise.
    The complete guide to growing berries and grapes / Louise Riotte ;
with illustrations by the author.
        p.   cm.
    Originally published: Charlotte, Vt. : Garden Way, 1974.
    Includes index.
    ISBN 0-87833-825-X
    1. Berries. 2. Viticulture. I. Title.
SB381.R55 1993
634.7—dc20   92-34629
                                                        CIP

Printed in the United States of America
10   9   8   7   6   5   4   3   2   1

✿ Contents

1.

Have Fun Growing Berries

WHO CAN? ANYBODY CAN

ven if you are a new gardener, by following a few simple rules you can be just as successful as an oldtimer who knows all the ropes. Small fruits are among the easiest plants for home gardeners to grow. In fact, I have wondered for years why more people don't realize how worthwhile these plants, bushes and vines can be both to the homesteader or the dweller in suburbia.

Sometimes people feel that they just don't have room for all the things they want to grow. And, if space at all is available, it often will go first to the vegetable garden, and small fruits are snubbed or entirely left out of the picture. I can understand this, for once upon a time, I felt that way myself. Then I began tucking in a few berry bushes and grape vines here and there, and I found they would grow just as diligently by the side of the garage, up and down the border of my front walk and on a trellis above the patio as they would in long, straight rows. Berries, many of which are members of the Rose family, are just as ornamental as flowering shrubs, as attractive in flower, delicious in fruit and delightful in their autumn suits of vivid foliage.

There are a lot of convincing reasons why home gardeners should grow small fruits, the most obvious of which is the economic. But, to my mind this is not the most important reason. The primary wisdom of growing your own is health. How else can you be sure that the fruit you are eating is completely clean, unsprayed with harmful chemicals or coated with preservatives? And the hands that picked the fruit are yours, or your children's, and no one else has touched it. Further, you are the boss in this, your own little organic world, and yours is the pleasurable decision of when to pick. You can take the Strawberries when they are a bright, glossy scarlet, the Raspberries when they are a deep red, black or gold and the Grapes at their blue-black best. You don't have to settle for half-green fruit, picked before it is ripe so it will stand up under lengthy transportation from farm to consumer. The span of time from plant to plate is, quite likely, less than five minutes.

And right here I want to give you my first helpful hint. You will, of course, keep your small fruits as clean as possible (and directions for mulching will be given in subsequent chapters), but if washing is necessary here is what to do. This is a secret I learned from a farmer's wife who always picked her berries at their juicy best.

"If you wash them when you first bring them into the house," she told me, "they are warm from the sun and the skins will break easily, lose juice and absorb water. Take the berries instead and place them in large freezer bags. Put them in the refrigerator for several hours or even overnight. Then take them out and wash them in a bowl of cold water. They will be firm and easily washed and you will lose almost no juice in the process."

I have tried this with many varieties of small fruits and find it works equally well with all of them. Handled this way they are especially appealing to the eye for serving whole. If they are to be used for jams or jellies you have more of the real product and lose little or none of the juice, which is often replaced by water when the berries are washed while still warm.

Another excellent reason for growing your own is the superiority of the product. Commercial varieties, understandably, are often grown for their shipping and keeping qualities, or even eye appeal, instead of flavor and digestibility. These reasons do not necessarily make them desirable. We know now that the very fact that some foods are easily perishable adds to the ease of their assimilation in our bodily processes. And easily-perishable varieties also are quite apt to be the most tasty and flavorsome.

As a gardener, on a country homestead, or even on a city lot, you can choose the very best varieties and pick them at their peak of perfection. Let your family enjoy them at their sun-ripened best, and know that you are putting an exceedingly clean and healthful product on your table.

Is it worthwhile to go to a little work and take the time to do this? You bet it is!

ARE PLANTS EXPENSIVE?

Strawberry plants, berry bushes and grape vines still are among the biggest bargains available. Understandably they have risen in price along with everything else, but so have the prices of the fruit you buy in the grocery store. And for the money you pay for a few pounds of Grapes or baskets of Strawberries you can still buy a number of plants and grow your own.

Moreover you don't need to buy a large number of plants to begin with. I always have believed that a variety of plants rather than a large number of anything was a good way to start. You will, of course, be the judge. If your family is particularly fond of Strawberries, for instance, you may wish to grow several varieties of these to the exclusion of everything else. But I like the idea of having many different kinds of small fruits ripening at different seasons.

Very little, if anything, need be spent on special equipment. If you make a vegetable garden it is quite likely that you already have the few tools you will need. Hoes, rakes, clippers and sprayers are all convenient, and if you do not have these, you may want to purchase them for efficiency and convenience. But, if you possess nothing more than a shovel and a sharp kitchen knife to start with, you can go right into berry production. Probably, you already own a garden hose which may be needed for watering your plants as the summer advances.

Consider the money you spend for plants and vines in the light of a permanent investment from which you will draw dividends for many years in the future. The dividends have a direct cash value, not only to your family for what you will save on fruit purchases, but in time you may even sell your fruit surplus. Surplus plants may also be sold or traded—or you might give them to a young couple just starting out garden-keeping.

ARE BERRIES AND GRAPES DIFFICULT TO GROW?

I am almost tempted to fib and tell you "yes," just so I can go on and prove to you how smart I am by knowing "how to." But the plain, honest truth is "NO." As I said in the beginning, anybody who wants to can. Some plants, especially those adapted to your particular climate, are so easy to grow that even brand new gardeners sometimes succeed by accident—like the fellow who threw some surplus Strawberry plants on his trash heap in the fall and found them the following spring growing and producing berries in a state of innocent glory without benefit of any attention at all.

Berries, as many of us remember from our childhood, have a great habit for growing and producing in the wild state, increasing their territory with the passing of the years, unsprayed, untended and constantly harvested by birds and animals. Here, where I live in Oklahoma, Dewberries grow wild everywhere in our state parks, around our many lakes and often all over the low-lying hills. Blackberries also have naturalized. And a small wild Grape, particularly delectable for wines and jellies, is very abundant in the fall of the year.

No, you won't have any problem growing small fruits, but my advice is to approach what you will grow with hard, practical common sense. If you want a carefree planting, where you will not have to work yourself to exhaustion to pamper things along, grow the fruits that are adaptable to your section of the country. If for instance you live in a cool, moist, northern section, Blueberries are for you. If you live in the warm, humid South, get with it on Boysenberries. On the other hand you can be assured of reasonable success growing Strawberries in just about any part of the United States. But here, again, you must know your berries. Certain varieties are more suitable than others for different sections. It will pay you to be particular and make your selections with care.

WHERE CAN THE NEW HOME GARDENER FIND SPACE?

This is a problem that stumps many would-be home grower of small fruits. But it need not be insurmountable. Just use your imagination, and you will find all sorts of nooks and crannies where you can place a few plants.

Boysenberries, for instance, can be trained to grow on a two-wire trellis. I have seen Strawberries growing as a border along a front walk, and mighty attractive they looked too, both in blossom and fruit. As runners developed, the gardener gently pushed those to be saved away from the walk and pegged them down where she wanted them to grow with large hairpins. From a few plants, purchased for the price of a box of berries, she thus began a plan of propagation which in time resulted in sufficient plants to go the entire length of the walk on both sides. The last I heard from her, she was planting Strawberries along her driveway.

Experienced gardeners will all tell you that berries prefer full sun, and so they do. But I learned long ago to temper rules with reason. If you live in a mild climate with ample moisture from natural rainfall you doubtless can follow this rule with good results. I happen to live in a dry, windy section. In the summer temperatures frequently go up past 100 degrees and stay there for a week or more during the daytime hours.

This can be pretty hard on Strawberry plants, which thrive best under conditions of ample moisture. I get around this by placing my strawberry rings (each equipped with its own sprinkling system) on the east side of my fruit trees. Here they receive completely adequate sunshine in the morning up to about noon. Then the shade of the trees gradually begins to spread its protective coolness over them as the afternoon advances, shielding them from the hot western sun of late day. This is when I turn on the sprinkling system, giving them a drink, but with plenty of time for the foliage to dry before nightfall.

Lots of times there is a small space between your garage and your neighbor's property line. Instead of letting this run to weeds, why not use it to advantage by planting it to bush berries? You will be pleasantly surprised how quickly some varieties will bend down their long canes and present you with new plants.

Do you have a service yard that needs a bit of screening? Here is another often-overlooked opportunity for growing grapes. These very obliging vines will grow at just about any height you want them to. Build your supports accordingly, and grow them on a wire fence or trellis. I assure you they will be far more attractive than a fence of plain boards, and you will have all the fun of picking your own grapes, too.

I have never found a hard and fast rule anywhere that says hedgerows must be of privet or flowering shrubs. If your climate is right for Blueberries they are most attractive grown in this manner. Or if you want an impenetrable hedge, consider Blackberries or some of the Ornamental Raspberries. When they attain their full potential, you will have both protection and delicious berries as well.

Even the tiniest home lot can find room for a Strawberry jar, barrel or ring. Some people even make raised strawberry beds out of old railroad ties so they can be comfortable and sit down when they pick the berries. If filled with good soil, this is a very practical plan, and who said anything had to be done the hard way? Any time I find an idea for growing something easily and conveniently I grab onto it, quick.

THE RIGHT KIND FOR YOUR SECTION

This is a matter on which at best, I can only generalize. It is very possible that you may be able to grow every kind of fruit which I propose to mention, even if you live in the far South or the far North. For I have observed there are "mini" climates—microclimates—in many places which enable careful gardeners to take advantage of or create conditions which will let them grow certain fruits which are thought to be "impossible."

Perhaps you are in a generally dry area but have some marshy land adjoining a small stream or lake. You may be able to take advantage of this and grow small fruits, such as Blueberries, which your neighbor only a short distance down the road could not.

Boysenberries, generally grown in the South, will succeed in the North where winters are not too severe. Two types of the Boysenberry were introduced commercially by Rudolph Boysen (Anaheim, California) in 1926: One with spines and one without, and often spoken of as thorny and thornless.

There is a difference of opinion regarding the two strains but most seem to feel that the thorny type is the hardiest. The thornless type also is subject to more serious rabbit injury during the winter. On the other hand the thornless is more easily trained on the trellis and more easily harvested.

Depending on where you live you may wish to try one or the other, for the dark wine-red berries are large, oblong in shape and of excellent quality. To grow them yourself offers a distinct advantage, for this berry is at its best when allowed to ripen on the vine, kissed by the sun to its fullest potential of sweetness and flavor. Because the berries are somewhat soft when fully ripe, commercial growers harvest them while they are still quite red (instead of purple-black) to insure their arrival on the market in firm condition, and so quality is sacrificed.

THE COUNTY AGENT IS YOUR FRIEND

For a ridiculous number of years I labored under the delusion that the County Agent would be of service only to the "big" farmer and rancher. Because I lived on a small city lot I thought I was precluded from requesting information. Nothing could be further from the truth. When I finally got up my courage and started calling and "dropping 'round," I found every courtesy and attention. You will find at your local County Agent's office, also, a number of small books and pamphlets on berries and grapes, most of which are free for the asking. Others will have a small charge to cover the cost of printing. Many of these are put out by the United States Department of Agriculture and additional copies are obtainable from the Superintendent of Documents, U. S. Government Printing Office, Washington, D.C. 20402.

Your County Agent will help you also with specific problems which you may be having because of climate, pests or diseases. If you are having troubles, quite likely others are too, and you can get a quick, timely answer which all the generalization I may make in this book would not give you.

If your County Agent is stumped (and he is not infallible), there still is your State Experiment Station to go to for answers. In fact you will probably find as I do that many of the pamphlets in your Agent's office are supplied by this source and deal specifically with the fruits and vegetables particularly adapted to your state or section.

I have before me now as I write, three such books originating at Stillwater, Oklahoma, put out by the Extension Service of Oklahoma State University. They are "Bramble Fruits," "Small Fruits For The Home Garden," "Growing Strawberries For Home Use," and a fourth book, "Grapes For Oklahoma."

All of these booklets, containing many pages of useful information, were sent upon request, and I'm quite sure the Experiment Station in any state you happen to live in will do the same for you. Each one of these booklets will give practical, down-to-earth information which will also include the names of specific varieties which have been definitely proven to do well in your part of the country. With the guesswork removed, you are well on your way to success in growing your berried treasures.

Don't hesitate either to ask advice of some of the older, experienced gardeners and small fruit growers in your area. I have always found gardeners to be the nicest, friendliest people imaginable, more than willing to share the information they have gleaned through many years' experience and often surplus plants as well. Since I've done this many times myself I know the glow of happy pride I get when someone asks me how I grow things.

Even now I'm making a list of friends who will be given extra Strawberry plants this fall when the runners now setting will be mature enough to move. Although spring planting of Strawberry plants is generally recommended, this is something we can do successfully in the South because of the long growing season. By spring they will be well established and far better able to withstand the heat and drought of summer.

While I agree that "it's not nice to fool Mother Nature," I don't see any harm in outwitting her just a little bit. Collect all the know-how you can from every source available, and quite probably you will succeed in doing just that.

In time, with knowledge and with a bit of extra care, you may even succeed in growing some of the "impossible" things, especially if you happen to have, as I do, a large bump of curiosity and a strong urge for experimentation.

2.

Planning Before Planting

eciding to grow some small fruits and getting started are two different things. By now you have probably walked around your homestead, or yard, and figured out a rough plan of "where to put."

You've been to your County Agent's office, talked to an experienced gardener or written to your State Experiment Station. Armed with this information, you have a pretty good idea of just which small fruits will succeed best in your part of the country. The next thing you will want to know is just where these varieties can be obtained.

If a local nursery or garden center cannot supply you—and no gift plants have been forthcoming—your next best bet is to secure a good supply of nursery catalogs. I have compiled in the back of this book a list of reliable nurseries with whom I have done business over many years. This list does not by any means include all the fine nurseries where plants are available. There are many others just as reliable. In a world where integrity has become increasingly difficult to come by, I have always found nurserymen unusually trustworthy people to deal with. Most of them lean over backwards being fair and honoring their guarantees.

So order your catalogs. Then sit down quietly and study the prices, descriptions and all the information they give you. But remember this: the beautiful pictures and glowing descriptions are of that particular fruit or berry at its best. Those grapes or berries undoubtedly were grown under ideal conditions, with the climate, soil and season exactly right for the product. Many new gardeners fail to take this into consideration, buy plants and plant them, and then wonder why their fruiting specimens do not even remotely resemble the marvelous pictures.

How do I know this? Because I went through this myself, being repeatedly disappointed in the results, blaming the nursery stock and everything and everybody except the real culprit—myself.

THE IMPORTANCE OF SOIL PREPARATION

Unless you are unusually fortunate—and few of us are—the soil around your homestead quite likely is far from ideal. It may even be subsoil and largely clay which has been dumped there as landfill.

A number of years ago when Carl, my husband, and I built our new home I couldn't wait to start all the planting I had been envisioning for the yard and garden. In vain he protested that there was a great deal of soil preparation which should be undertaken. I stubbornly refused to listen. That first summer the soil—without benefit of organic matter or mulching material—baked hard and dry and most of my weary, discouraged little plants gave up the ghost.

At last I got it through my hard head that things just weren't done that way, and in the fall we started seriously to gather materials for our first compost heap. Bermuda grass, which grows easily in our climate, had not been difficult to establish on the lawn and we saved the clippings. Leaves gathered in the fall from our large hackberry trees and other sources were added to this. Manure and strawy stable cleanings were secured from friends who owned ranches. All of these materials were layered up along with kitchen refuse and a sprinkling of agricultural lime and covered with soil. For neatness such a heap can be put in a box with one side open for easy turning, or built of cement blocks.

Of course this was just a beginning, and the process of gathering materials for composting has gone on continuously ever since. But by spring we did have enough to enable us to plant a small vegetable

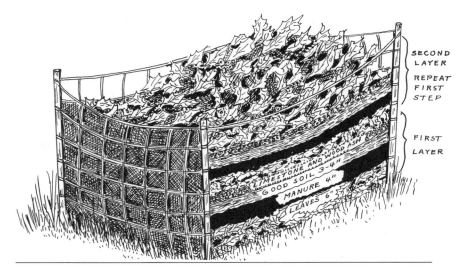

SECOND
LAYER

REPEAT
FIRST
STEP

FIRST
LAYER

LIMESTONE AND WOOD ASH 3-4"
GOOD SOIL 4"
MANURE 6"
LEAVES 6"

You, too, can make compost!

garden with fair success. As time went on and the fertility of our garden soil improved we gradually acquired a surplus of compost, and it was then that I started thinking again of small fruits.

By now we had more land, having first taken an option on and then purchased the lot which adjoins our home place on the south. Though by no means approaching the proportions of a small farm, we are fortunate in having a bit more land than the average city home can lay claim to.

Also by now we were the proud possessors of a small but efficient rotary tiller. Though a tiller is not necessary to berry or grape production on a small scale, we found it greatly improved the tilth of our garden soil. We used it for the flower beds, to plow along the walks, by the sides of the driveway and garage and even to break ground for our grape arbor. On a large acreage or a suburban homestead a tiller really will cut down on the hand labor.

Once all this was accomplished, adding organic matter to the soil became a routine part of our life. And the next time around, when I purchased my strawberry plants, berry bushes and grape vines, I felt reasonably certain of success in growing them—as I was.

Let me emphasize here that in small fruit growing you never should add fresh manure to the soil. It should always be well rotted, preferably through a composting process. Otherwise, added to the soil it will cause a temporary deficiency of that most essential plant nutrient, nitrogen.

If you cannot obtain fresh manure and compost it yourself—and I will admit it is getting harder and harder to come by—you can buy dried cow manure. This dried product, bagged and odorless, is obtainable at most garden centers and has already gone through the composting process.

Manure, in my opinion, still is just about the best soil conditioner you can use. It not only provides the food elements that plants need but it will also improve the soil texture, making it spongy so it can absorb and hold water. Thus the abundant supply of moisture that small fruits require is readily available to them.

The soil texture is so loosened, too, that excess water can drain through it. This is necessary, for if the roots of your plants sit in water for any length of time they will die or be seriously retarded in growth. They do not like wet feet. No matter what type of soil you have, sand or clay hardpan, you can improve it by mixing in well-decomposed manure. While I know we live in a time when "instant" everything is in demand, one of the first rules a gardener must learn, as I had to do the hard way, is patience.

If the ground in which you plan to plant your small fruits has been in sod, it is always a good idea to plow it up a year or two in advance and grow flowers or vegetables in it first. The reason for this is to get rid of those nasty small enemies of fruit production called white grubs. The soil under grass quite often is well populated with these little mischief-makers, which are exceedingly destructive to the roots of small fruits.

While manure will add that most necessary ingredient called humus to your soil, there are other organic possibilities, so if manure is not easily obtainable don't give up. You can use rotted leaves, known as leaf mold, (sometimes your city street department will bring these to you free of charge if you put in an early request). This is the natural food of berries that grow in the wild. Peat is good and so is green manure. This term "green manure" used to puzzle me greatly, but it is achieved by nothing more mysterious than the sowing of winter rye in the fall of the year. When it is about eight or nine inches high it should be spaded or plowed under. Moisture from spring rains usually will cause it to quickly decay. As this takes place, it will greatly improve the texture and nutrients of your soil.

Most gardening books seem to be written for the gardener who has room enough to do everything on a grand scale. If you have lots of room you are lucky, but if you are one of those who has space for just a few plants you can create a "mini" soil environment for each plant you set, making the most of the manure or compost that you have by concentrating it where you need it most, instead of spreading it over a wide area.

SETTING PLANTS

Soil requirements and planting methods for Blueberries and Straw-berries will be set out specifically later, but for most other small fruits and grape vines here is a good way to set individual plants:

Dig a hole about the size of a bushel basket, deepening it to about three feet. Be sure to put aside and save your topsoil which, if you are lucky, may be eight or nine inches deep. This topsoil is characterized by a richer, darker look, especially when it is moist. The subsoil generally is lighter colored (indicating that it is devoid of organic matter), and of a clayey, sticky texture.

This subsoil is largely worthless for plant growth. Unless you can use it to fill in a depressed area somewhere else, remove it entirely.

Your next step is to provide for drainage. Remember that hole is still surrounded on all sides by packed, sticky subsoil. If drainage is not provided it can become a small pool to collect water.

Begin by throwing in a two or three-inch layer of coarse gravel (you may be able to get this from a lumber yard or a cement company). Top this with several more inches of gravelly sand or finer gravel.

Now you have a choice. You can either start mixing that topsoil you saved with your compost, decayed manure, or whatever organic material is at hand, or you can start layering your materials in the hole. If you want to mix first, a wheelbarrow is a handy tool for the job. Use a hoe and just keep mixing until the color of the soil mix is uniform.

Here is what you do if you want to use the layering method:

Place a three-inch layer of topsoil in the hole over the layer of fine gravel. Then add a three inch layer of manure or compost. Continue building up the layers, sandwich-like, as filling the hole progresses.

If you use dried manure it will probably be more concentrated than that which you have composted yourself with other materials incorporated in your own mix. For this reason it usually should be added more sparingly. Since brands differ, be sure to read the directions on the bag and act accordingly.

Remember also that manure will not provide two other important plant nutrients, phosphate and potash, in which your soil may be deficient. A soil test (which can be made by taking a soil sample and sending it to your state Experiment Station), definitely would determine this. However, for such a small-scale soil preparation as we are now doing you can provide for this possible lack by adding a cupful or two of bonemeal to the manure and topsoil. Bonemeal is nothing more or less than ground up animal bones, and as it decays both phosphrous and potassium become slowly available to the plants.

When all the layers are in, use your spade or spading fork to mix the ingredients thoroughly.

The hole should be filled a few inches above ground level, for your "fluffed up" soil will settle. Make a depression in the center and gradually add water until the mixture is thoroughly saturated. Let it sit for several weeks before planting your berry bushes or grape vines in it.

Depending on where you live, you may even find it more practical to do your soil preparation in the fall of the year, so that all is ready for early spring planting. A late spring with frozen ground may add to your gardening problems, delaying your chances of doing this early enough so that it will be ready when your nursery stock arrives. If the planting sites have been prepared in advance and the surface well mulched, the chances are that, without any problem, you will be able to dig the necessary hole, large enough to receive each plant and to hold the roots without crowding them.

Just one more word of warning here. When you plant, leave well enough alone and don't add any commercial fertilizer or even well-decayed manure around the plant roots. It could be fatal. The reason for this is that the salts of the fertilizer will add to the strength of the soil water, actually causing it to draw moisture from the roots instead of penetrating them. This condition is called burning and usually will kill the plant.

I do not advocate adding any fertilizer at all for the first year to that previously dug into the soil. Plants should have a chance to get over the shock of transplanting, and the addition of more fertilizer is quite likely to be a hindrance rather than a help.

As the plant continues to make good growth, you can place more compost and rotted manure in a circle around it and work it lightly into the soil. This will benefit the feeder roots which are close to the surface. If you have placed mulch around your plants, draw it back and replace it after this operation. Never cultivate deeply or you will destroy many of the feeder roots and do more harm than good.

TAKE CARE WHEN YOUR STOCK ARRIVES

Probably more small plants lose their lives through inattention on arrival than for any other reason. If you can buy the varieties you want freshly dug from a local nursery you are indeed fortunate. Make the most of this, and get your stock into the ground as quickly as possible when you arrive home.

If you cannot buy locally, make your catalog selections early. If you can, buy from a nursery in your area which stocks varieties suitable to your climate.

Here is what to do when your stock arrives after a trip through the mail: Soak the roots of your berry bushes and Grape vines (not Strawberries), in a bucket of water for several hours. If the stock appears to be very dry you may even lay it in the bathtub for an hour or two and completely cover it with water.

When you plant berries and vines, never expose the roots to air or sun more than is absolutely necessary. Keep the roots covered with wet burlap if you have it. If not, insert them wet in a large plastic bag and carry this along as you work.

In the actual setting out of plants the most important thing to remember is this: the soil should always be packed firmly around the roots. If it is not well firmed, air pockets will be left around them. The roots then cannot absorb moisture and the plant will die.

Most stock should be set one inch deeper than it stood in the nursery row—this line will usually be apparent to the eye. Always spread the roots out carefully. Trim off broken roots and tips of roots with a sharp knife.

Once the plant is set, tamp the soil by walking around it and packing the soil with your feet. Step hard and pack firmly. The lighter your soil, the harder you must tamp.

Slowly pour water in to fill the hole. Let it settle and then fill with earth and tamp down again. Leave a depression around the bush or vine to collect and hold moisture. (Directions for Strawberries will be given in the next chapter.)

The final step is either to loosen the top few inches of surface soil to avoid cracks or to cover the ground with a good mulch. Either method will prevent rapid drying, but personally I prefer to mulch.

For many reasons it is not always convenient to plant nursery stock immediately. If this happens, the stock should be "heeled in." This simply means digging a trench in an out-of-the-way place, putting the stock in it, and covering the roots with soil at least six inches deep. Water to eliminate air pockets. If weather is particularly warm, dig your trench in a shady place. If the stock is to remain in the trench for any length of time, I like to place a mulch over it to prevent further drying out of the roots.

Always set the plants close together in the trench. Work moist earth around the roots and pack it well. Berries and vines can remain like this for a reasonable length of time, but get them into their permanent location before they have ended their dormant period. When their buds begin to swell they have started growing, and don't let this happen. Get them out at least several weeks before leafing-out time.

While Strawberries may be heeled in as a temporary measure— and if you have many plants this may be the most convenient way to handle them—there is another method advocated by a leading nursery, and which I have tried and like.

You can do this if your Strawberries arrive either in hot or cold weather and planting has to be delayed. Place in a polyethylene bag

To heel in plants, place them in a trench with the crowns at ground level. Pack soil firmly around the roots to prevent air pockets and keep the soil moist until ready to plant.

(one of the large freezer-type bags will do), and store them in the 50-degree section of your refrigerator. Remember I said refrigerator—not freezer. They will keep just fine like this for a short period of time, but plant them also as soon as you can.

Just so we can cover all bases, there's one more thing to mention. I've told you to unpack your nursery stock as soon as it arrives and plunge the roots into water, but there is an exception to this rule. Sometimes it happens that the weather is very cold when your plants arrive, and perhaps they have been hauled for a considerable time in the back of a delivery truck. If you suspect that the box may contain frost, do not open it. Instead do this: Store it in a cool place but one where the temperature is above freezing. If the plants are allowed to thaw out slowly, over a period of six or seven days it is quite possible that they will be completely undamaged. But it you thaw them out quickly in a warm room they may die.

All of these directions sound as if small fruit plants were simply passive things like bags of flour or a pair of shoes, but they aren't. That plant, bush or vine will help itself all it can. It is a small bundle of determination-to-grow. It wants to live. It is young and full of energy. It will breathe, sleep, eat and drink. To do this well it must be in a congenial environment. This agreeable climate must include light and shade, sufficient moisture and soil to its liking.

If your garden does not possess these things already in the right quantity, then to be successful you must provide them so far as possible. Place your plants where they will receive sun light, give them food when they need it and plenty of water when they are thirsty. Give them every opportunity, and they will surprise and delight you by their cooperation.

Talk to your plants as you go about the pleasant task of placing them in the earth. Encourage and praise their efforts as you weed and water and mulch. Does this really help? I think it does.

Now let's help ourselves to Strawberries, for many the best loved of all small fruits. And they're so easy to grow that you can do it with your green thumb tied behind you!

3.

Sun-Loving Strawberries

 trawberries belong in every home garden. They're not fussy about soil or climate and will thrive in all sections of the United States, including Alaska. The two things they do require, good drainage and full sun, can be provided in any section of the country. If drainage is a problem, grow them in a raised bed. Six inches will be completely adequate, for they do not root deeply. Ample sun is necessary but, as previously mentioned, I find that in an extremely hot, dry climate they do better when they have afternoon shade.

Unquestionably the size and quality of the crop will also be influenced by deeply enriched soil. Peatmoss is particularly helpful on sandy soils because Strawberries, while they do not like wet feet, need plentiful moisture to grow juicy and flavorful. In a warm climate like mine they do especially well when abundant moisture can be supplied during dry weather. Ample water is also most important for new plants when spring set, so that they may become well established before summer. Unless strong crowns are formed a good crop of berries will not result.

Strawberries were not cultivated until the fifteenth century, although they had grown wild over much of Europe. A wild species was also found in America by early colonists. The American berries had a better taste than the European ones, but were very small and sometimes dry. A Frenchman, Captain Frezier, arrived in France in 1712 with a few Strawberry plants that he had obtained from Chile. These produced big, tasty Strawberries, which were later to be grown in both France and England. By accident, some were grown near plants from eastern North America and the exchange of pollen produced excellent crops of large berries.

UNDERSTAND YOUR STRAWBERRIES

In southern regions such as Florida and the Gulf Coast, Strawberries often are grown as annuals. Plants are obtained and set out from January on through April. Usually they start into growth immediately and bloom and set fruit before forming runners.

In June or July, runners are taken from the parents, planted in other beds, and they in turn produce daughter plants. The daughter plants then are used for planting from September until mid-December.

In most regions of the United States, however, Strawberries are treated as perennials, which are plants that live for several years, blossom and bear fruit each season. The Strawberry usually bears its best crops the first two years, but sometimes under ideal conditions plants will last a year or two longer. But most varieties, if held beyond the two-year period, will produce berries that are progressively smaller and less tasty. For this reason a program of plant replacement is desirable.

Common name: Strawberry

Botanical name: *Fragaria*

Days to maturity: Plants are set in spring or fall to crop the following spring and summer.

Soil: Well-drained, slightly acid.

Nutrients: Lots of humus. Light application of K sources; medium application of N and P.

Water: 1 to 2 inches per week; moist soil, but never soggy.

Light: Full sun.

Spacing: From 1 to 2 feet apart, in rows 4 to 5 feet apart.

When to plant: Set transplants as soon as soil can be worked in spring. In the South, transplant in late summer and fall as soon as hot weather has passed.

—*Organic Gardening*

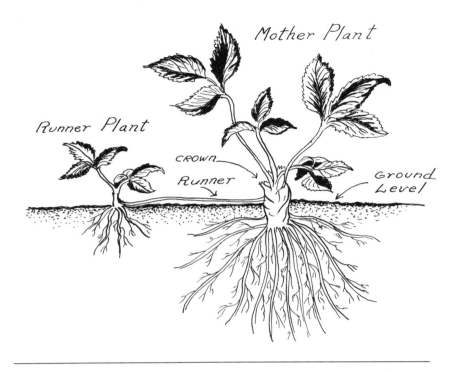

Strawberry plants produce runners that then take root and form new plants.

Propagation of Strawberries is accomplished by means of runners. These simply are stems, sometimes several, which grow from the original plant and "run" over the ground. They will root easily in loose soil, forming daughter plants. By means of a hairpin or bent wire you can peg them down where you want them to grow.

This ability to root at a slight distance from the parent plant is a form of natural layering. Leave the plant until it has formed a good root system. It may then be cut loose and allowed to grow or you may dig it up and plant it elsewhere. Once you have purchased your plants and made your first bed, you can go on propagating your Strawberries indefinitely.

There are a number of variations on the Strawberry theme, some bearing early, others mid-season or late. But Strawberries usually are spoken of as "June bearing" or "everbearing."

You can lengthen the picking time of the spring-bearing types by selecting several different varieties that will fruit progressively as the season advances. But many people prefer berries which set all their crop within a relatively short time, enabling them to pick the crop for canning, freezing, jellies and jams all at one time and thus complete the task quickly.

The spring-bearing types set their blossoms profusely in the spring and the pretty white flowers soon set fruit, turning quickly into berries, green at first and then white and gradually turning to brilliant, shining scarlet under the kiss of the warm sun.

I would describe the everbearing type more accurately as "repeat bearing." During their second year they may be expected to bear two good crops. Set out in the spring, within a short time they will start setting a crop of fruit. The plant then will rest, and toward midsummer will bloom again, setting a second crop in early fall.

There are distinct advantages in growing both types, and a true Strawberry addict probably will want some plants of each. I grow both and find the fall crops decidedly welcome.

One of the best reasons for growing the repeat bearers is in having at least a small crop the first season. Most, though not all, experienced gardeners advise removal of all blossoms that appear during the first season on the spring-bearing varieties. This gives the plant a chance to become well established, throwing its strength into growth rather than into the production of berries. I feel that this applies to everbearing types also, but in a more limited way. When the plants are small and just starting to grow I pinch off the first blossoms, allowing the plant to set fruit when its size and strength increases.

PLANTING AND TRAINING

How many plants should you buy? There is no ready answer to this question, as space and your own inclination to care for the planting will enter into the picture. But if you have room for only a strawberry ring, six feet in diameter, this will accommodate fifty plants nicely. Two such rings would assure ample Strawberries for a family of four. You could plan on buying enough Strawberries for one ring the first year and then set the second one with runner plants as they form.

Even two dozen plants well cared for and set in good soil beside your front walk or along your drive will give you a good return. Most people will want more than this, and, necessity being the mother of invention, probably will find a place to put them.

On the other hand, if you have plenty of space you may want to set out several hundred plants. Remember no Strawberry ever will need to go unloved or unwanted, for unlike many other fruits every bit of the crop is usable. If you have more than you want for eating out of hand, for making shortcake or slicing over ice cream, you may freeze the balance easily and preserve them for winter treats.

Strawberries usually are sold as bareroot plants, being shipped in bundles of as few as twelve or as many as fifty. When they arrive, open the bundle at once even though you may not be able to plant until some time later.

Heel the plants in or store them as previously directed in your refrigerator. Before setting your plants, trim the roots back so that they are about three or four inches long, also removing any old leaves or blossom buds. Leave only the new, bright green inside leaves.

If possible, the actual setting of the plants should be done when the soil is moist. I like to open the hole with a small hand trowel, one that is wide enough to allow the roots to be spread out fanwise. Then I cover them and press them in firmly, being careful not to get soil into the centers of the crowns.

The plants should be set so that the crown is at ground level. Below ground level it will smother and growth will be retarded. If set above ground level the exposed roots either will die or be injured seriously.

Strawberry plants have a root system consisting of many fibers, and often it is all too easy to set the plants loosely, even though we know better and are making an effort to set them correctly. That's why I like the broad blade trowel. A rounded instrument which makes a small hole will not permit me to shake and spread the roots. Crowded together, the roots will not make good contact with the soil and cannot make their best growth.

One person can make a small planting easily if working alone, but for a large planting two people can get the job done far more quickly and efficiently. One can open the holes and the other can carry the

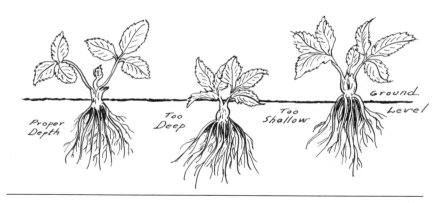

Plant strawberries at the proper depth. Spread roots fanwise. Crowns should just rest on the surface (center above). Trim roots to 4 inches. Pinch off blossom buds and old leaves.

plants in a bucket of water, dropping them in as the spaces are made ready to receive them. As this is done the soil should be pressed with your foot to avoid air pockets. If the soil is dry the plants should be watered also. For many plants it may also be advisable to use a spade instead of a trowel to save frequent bending over. The method is exactly the same and much time is saved.

Having set the plants, weeds must be kept down either by frequent cultivation or by mulching, which also will help to conserve moisture, as well as provide a favorable environment for the runners to root and establish themselves. This they will do much more quickly in a loose or mulched soil than in one that is hard and packed. A soil well-supplied with organic matter, either in the form of compost or well-decayed manure, is also worked more easily and is less subject to drought.

Remember that while Strawberries will grow on a wide range of soils, providing they are well supplied with organic matter, the soil texture actually is secondary in importance to good drainage and an ample supply of moisture.

The ideal soil for Strawberries is a deep sandy loam, overlying a subsoil that is well drained but retentive of moisture. Good surface drainage also is important because Strawberry roots are easily injured if they remain for any length of time in a saturated soil. Such soils, too, may heave during the winter months and cause great damage to the roots.

If you can plant your Strawberries on a gentle slope having a well-drained, porous subsoil, you are well on your road to success. Good air circulation also is a desirable factor. It gives at least a partial protection against frost, which may gather in low spots surrounded by higher ground. Fungus diseases are less apt to attack your plants, also, where there is good circulation. Most Strawberries are planted in the spring, especially in the North—March, April or May generally being favored. On the other hand if you live in the South, fall planting or even early winter is the best time.

How far apart should plants be spaced? The method you choose to grow your plant is the determining factor here, for they may be trained to grow in several different ways.

PROPAGATION—MOTHER AND DAUGHTER PLANTS

Mother plants give birth to children by themselves, but the training of these daughters is largely your decision to make. The slender runner stems, as they come in contact with the soil under suitable conditions, will take root at a node (a joint in a stem), commonly the third, and form new plants.

Plants growing in soil which is either naturally rich in humus or to which organic fertilizer in the form of compost or well-decayed manure has been applied, will produce more runners than plants treated with inorganic fertilizer.

The best fertilizer to use if you would stimulate plant production is bone meal and fish meal. Ample moisture also is beneficial in causing prolific plant production.

Here is a method I often use for rooting Strawberry runners. It may be used for Strawberries on flat ground, but I find it particularly good for rooting the runners of my pyramids.

I begin by saving waxed-paper quart milk cartons, washing them well and setting them aside so they will be ready when I need them. When I have a sufficient number, I open the carton tops, press the cartons flat and cut them in halves with a pair of scissors. I use both tops and bottoms, simply reclosing the tops as they were originally. Next I make openings for drainage near the bottom of each half, on opposite sides.

Next I take a small wooden box or flat and place the cartons in it, filling each one with a good soil mix, about half compost and half garden loam. The box then is immersed in water (I use my kitchen sink for this), and left until the soil is completely saturated. Then it is removed and allowed to drain.

Then I take the box and cartons to my strawberry pyramid where the plants are putting out runners, often dangling around the edges of the rings with nowhere to go. I place one of the containers at the end of each runner where a new plant is forming, pegging it down in the carton with a piece of wire or a large hairpin to anchor it in place.

When roots have become established on the new plant I cut the daughter plant free of the mother and my young Strawberry is ready to go to a new home.

In anticipation of starting a new bed I have the soil already prepared and holes already opened up large enough to receive the

Strawberries by the Stars

If by chance you are a practicing astrologer, Strawberries should be planted in the third phase of the moon, preferably in the sign of Cancer, Scorpio or Pisces. Other berries are best planted in the second phase under the same signs.

Cultivate to destroy weeds when the moon is in a barren sign—Aries, Gemini, Leo, Virgo or Aquarius, and decreasing—fourth quarter preferred. Apply organic fertilizer during the decrease of the Moon, third and fourth quarters. Harvest fruit in the decrease of the Moon and in dry signs. Irrigate when the Moon is in a watery sign. Prune during the decrease, third quarter and Scorpio.

transplants. As I work I immerse each carton in a bucket of water for a few seconds and the plant and soil slides out easily when held upside down. Hold the plant carefully when you do this so it will not fall on its little head.

Planting a new bed this way is accomplished quickly and easily, with no shock to the transplants and no growing time is lost. Water it of course when the job is completed. If you are able to obtain peat pots they may be used instead of the milk cartons and pot and all planted in the new location. If they are quite dry, soak the peat pots in water before you plant them in the new site.

This method works best with an overhead irrigation, which is provided by the sprinkler system when strawberries are grown in the three-tier pyramids, or rings. The soil in the cartons receives water right along with the rest of the bed. If used on a flat bed often it is advisable to bury the carton partially in the soil to help conserve moisture.

Transplanting runners this way is especially good for fall planting in southern sections, enabling the newly established plants to make good growth before colder weather.

TO PINCH OR NOT TO PINCH

I always thought Strawberries acquired their name because they were most commonly mulched with straw, but this seems to be wrong.

The Strawberry plant is a member of the Rose family a fact reflected by its lovely, scented white blossom that resembles a wild Rose. It grows close to the ground, its leaves in groups of three, growing on a short woody stem. The berries like the blossoms have a delicate, pleasant fragrance—and we all know their delicious taste, which is especially welcome early in the year.

It is these berries which gave the plant its early name of "Strewberry," for they seemed to be strewn among the leaves of the plant. In time the name changed to Strawberry.

When pinching the first blossoms from your Strawberry plants (to prevent them from setting fruit), the stems on which the blossoms appear should be removed entirely by snipping or pinching them off at the base of the plant. It often is difficult to bring yourself to do this, but the rewards are great. The strength of the plant will not be depleted by fruit setting too soon, and the resulting crop, when the plant has matured enough to set it, will be far more generous both in quantity and quality.

Not every Strawberry flower will produce fruit. Some flowers do not have stamens. These must be planted near plants that have stamens, so that their pollen (pollen corresponds to the male sperm in animals) can fertilize the female parts. This type of pollination is known as cross-pollination. It may be accomplished by the wind or by insects.

HOW LONG SHOULD A PLANTING LAST?

This is a debatable question since so many factors enter into the answer. Fertility of soil, conditions of moisture and health and vigor of the plants all determine the life of a planting. Two to four years generally is considered the length of time a planting should be maintained, but sometimes you can leave strong plants for five or six years in a home garden. Commercial growers seldom maintain a planting for more than two or three years at most.

Here is a way that a friend of mine handles his Strawberry bed to prolong its life. He maintains that Strawberries are very hardy little plants, and the results he achieves seem to bear out his theory.

In early summer when his spring-bearing Strawberries have stopped producing, he runs his power mower (set high) over the entire bed. This cuts the tops and old leaves off the plants, also cuts down any weeds. He says that his Strawberries are tired after the bearing season and need to recuperate. He believes that cutting off the tops will give the root system a chance to rest with no necessity for feeding and watering all the old leaves.

With the old leaves chopped off, the plants can give all their strength to producing new leaves and fruit buds for the next season's crop. To do this effectively he sets his mower at a two inch height so the crowns of the plants will not be damaged.

Next he takes his rotary tiller and turns under every other row of the bed, tilling under four rows and leaving four rows, alternating these rows each year. The plants that are left send out runners and soon fill the cleared rows. This means that when the old rows are turned under each year, he has new plants every two years over the whole patch.

Without a feeding program, he soon would find his soil depleted of organic matter, so he takes advantage of the bare rows to turn under compost or well-decayed manure, also mixing in either bone meal or phosphate rock. Keeping the soil enriched keeps production high and his bed never seems to run out.

He also tells me that weeding is much easier after the tops of the old plants have been cut off. And once the remaining rows have been well cleaned, he starts mulching to help conserve moisture during the dry summer months. Also, until the new runners start to take over, he goes over the bare rows about once a week, keeping the soil loosened, so that the runners can dig in easily and start forming roots.

CARE AND FERTILIZATION

The first year after starting a new bed, the most important task is to keep the soil well cultivated. If you use a rotary tiller for this job set it so that only the top few inches will be stirred. If done by hand, use a rake or a hoe, remembering that Strawberries have shallow roots and you must not go too deep. In reasonably good soil with adequate moisture, your young plants are full of vigor and should grow out well.

Have the Kids Outgrown the Sandbox?
Convert it into a Strawberry bed! Plant everbearing Strawberries directly into the sand. Adding fertilizer (preferably well-decomposed cow manure), spring and fall will keep the plants going and prospering in the sharply drained soil. Few weeds appear and these are easily removed. The plants continue bearing for several years.

Attend to this matter of cultivation at least once every two weeks, so that the soil will remain loose and pulverized. Not only will this keep down weeds, but it will also help to conserve moisture. Never go deeper than an inch or two, and be careful not to throw dirt over the crowns of the plants.

If the soil has been well prepared, additional fertilizer usually will not be needed during the first growing season. By the time the fall of the year rolls around you will probably have additional compost to spend on the berries. Put a circle of this around each plant and work it lightly into the soil.

MULCHING

Mulching is important for several reasons, and the type of material (whether straw or salt hay) is not in my opinion terribly important. You will probably use whatever you are able to obtain. Here in Oklahoma I use prairie hay, scattering it around and under the berries in February. If I do this early the bloom stems will come up through the hay and the berries, resting on the mulch, will be quite clean at harvest time. If you don't mulch, your berries still may grow out well, but the resulting sand and grit will make you regret your oversight.

Mulching in the fall, especially in northern sections of the country, is important for another reason—winter protection. The crowns of the plants which remain unprotected may be severely injured by low winter temperatures. This will be apparent the following spring, evidenced by reduced growth, reddish foliage and eventually when the berries ripen by the wilting and possible collapse of the plants. Mulching also will prevent the plants from being heaved out of the ground by alternate freezing and thawing of the earth.

To prevent these small tragedies, mulch should be applied before temperatures drop below twenty degrees, but after the ground freezes in the fall. Cover the plants to a depth of two or three inches with the mulch. Leaves, straw, sawdust, even pine needles may be used. In the spring part of the mulch should be raked off, leaving enough so the berries will be clean and to conserve moisture and keep down weeds. Sawdust tends to acidify the soil so use it sparingly.

Here in the Southwest, our winters are not severe enough to cause heaving, but I still like to put on a light mulch to retain moisture and check the weeds. When most of this has disintegrated by spring I start putting on the prairie hay to keep the berries clean and to prevent the ground from drying out due to our high winds.

If you have tended well to the mulching of your plants you will have even less work during the second season than the first. The mulch will eliminate the need for cultivation, and any weeds that appear can be pulled by hand easily.

This will be your year for a real harvest, and hopefully it will be a "berry good year." Remove no blossoms; let them form at will. Welcome every one, for the more the merrier. You should have a full crop from the spring-bearers and two crops from your everbearers.

If you have all the Strawberries you want from the spring-bearers alone, here is another idea. Keep the spring blossoms pulled from the everbearers with an eye to harvesting a larger crop in the fall.

After the everbearers have borne this fall crop, do not try to hold the plants for another season. Having already borne three good crops they are exhausted. They should be dug up and replaced with new plants.

Any time you cut or shear off leaves or dig up plants, remove them from the bed entirely. This will help to prevent the spread of any disease. Burn the foliage if possible or remove it from the premises.

HONEYBEES AND BIRDS

Perhaps because I once kept several colonies of honeybees I am especially fond of the little rascals. And this love for the small, busy creatures is one of the reasons I never spray my Strawberries with harmful chemicals of any kind. To watch them diligently going about their work on the Strawberry blossoms is one of my greatest garden joys.

Honeybees aid in the pollination not only of my Strawberries, but my other berries as well, doing their thing, as they whiz about on blackberries, boysenberries and grapes. They also pollinate blueberries, gooseberries and cranberries.

Birds, too, are friends, but especially during the fruiting season must be prevented from doing the harvesting before I can. Unerringly they will find the bright red cheek of my Strawberries, often before I do, and have themselves a small feast.

There are several ways to prevent this. Most Strawberry pyramids may be purchased with frames over which netting can be spread, the mesh open enough to admit sun, air and moisture, but preventing the birds from entering. If this is not feasible where you have larger beds, try cutting garden hose, no longer usable, in two- to four-foot lengths. Laid in your Strawberry beds and curled around a bit, these will resemble snakes, and birds are afraid of snakes. Move them occasionally from one place to another.

There are various bird scaring devices on the market such as flying disks, made of shiny aluminum which spin and twirl in the breeze. They operate on the principle that animals and birds fear the unknown and unfamiliar, and that the reflections and crackling sounds will frighten them away. I find that these disks do work for a short time, but also should be moved about.

INSECT PESTS AND DISEASES

The best way to prevent disease is to purchase virus-free plants. This is not as difficult as it was once, since practically all stock offered by present-day nursery companies is guaranteed virus-free. The symptoms of virus are not always clear cut and easily apparent. It is manifested mostly by weakened and devitalized plants that grow poorly and bear scanty crops or none at all. If virus seems to be present, sometimes it is best to destroy the plants and start over again in a new location.

I already have said that Strawberries never should be planted in land which just previously has been in sod, since it frequently harbors white grubs. Use such land for something else for a year or two. If you are troubled with cutworms, there are several things you can try: Strew crushed eggshells around your plants and cover the shells with soil. Strewing old rusty nails also may be of help. Steamed bonemeal will keep leaf rollers away.

Root rot sometimes is caused by over-fertilization. It also may be caused by the use of fresh manure, so use only that which is well-decomposed. **Powdery mildew**, which affects some varieties, may be controlled by dusting with wettable sulphur.

One commercial Strawberry grower insists that he has no **botrytis** or **brown rot** disease because he uses dolomitic limestone and colloidal phosphate in his fields, plowing this under in early spring. He also believes that the addition of dolomite results in a firmer berry. Since this man also leaves his plants in for four years—three of berry production, which is unusually long for a commercial producer—it would seem he has something.

The important Strawberry diseases are **gray mold** (which kills the blossoms and rots the fruit), **red stele** and **Verticillium wilt**, both carried by soil-borne fungi that infect the roots and kill the plant. Leaf spots and blights are symptoms of some diseases. Since Verticillium wilt is a soil-borne disease which enters through the roots, avoid planting Strawberries after other crops that are susceptible to this disease—such as tomatoes, potatoes, eggplant, melons, beets, peas, brambles and roses.

Your Strawberries should be extremely successful if a small number of precautions are taken against pests. The most damaging insects are the **strawberry weevils**, which are called "clippers." This is because they kill the buds and fruit by severing the stems, leaving them hanging as though partially broken off. The cat-facing (the injury resembles a cat face), sucking insects produce a deformed, low-quality berry. The **crown borer**, a white thick-bodied grub, one-fifth of an inch long, feeds inside the plant at the soil level so that the plant dies or is seriously weakened. To control the crown borer, locate new beds more than 300 feet from the old beds and plow up the infested patch immediately after harvest. **Sawflies** are controlled by white hellebore.

Good cultural and sanitary practices are very important in controlling insects and diseases. Other things to do are:

* Plant Strawberries on land that has been under cultivation for at least two years.
* When possible choose varieties that are resistant to disease.
* Work old beds over immediately after each harvest.
* Obtain virus-free plants for new plantings. Do not try to use your own plants if they have become infected.

Since certain pests and diseases often are prevalent in a specific area, while other places are relatively untroubled, again I suggest that if you have a problem you should call on your County Agent or your State Experiment Station for advice.

WAYS TO GROW STRAWBERRIES

As Borders For Walks And Driveways

Low-growing, Strawberries are ideal for use as a border for a walk. Their attractiveness both in flower and fruit make them just as lovely to use as many perennial flowers are for this purpose, and their leaves still are pretty when the fruiting season is past. If you have no other place to grow them, consider a border along your front walk, garden walk or driveway, or use them as a border for your flower beds.

The Matted Row System

This is the easiest way of all to grow Strawberries and, for effort expended, will give you the largest number of berries. Simply allow mother plants to set runners at will until you have a solid bed of plants as wide as your space permits.

Under this system the fruit will be abundant but smaller than what is produced by the hill system. But if you do want matted rows, start by spacing your plants two feet apart in rows three feet apart. Then let them all grow together.

Some gardeners prefer a variation of this method called the **hedgerow system**. Here only certain runners are allowed to set plants, all others being cut off. To use this method, place original plants about

Do not plant Strawberries where vegetable members of the nightshade family (tomatoes, potatoes, peppers, eggplants) have grown the previous season. These crops frequently have a buildup of verticillium wilt in the soil.

two feet apart in rows three feet apart. Let each mother plant form only two daughters, one on each side so that they form straight lines up and down the row.

A row of plants formed along each side of the middle row will result in what is called a **double hedgerow**. Good results are obtained with this system and berries are larger, but the amount of hand labor will be increased.

The Hill System

This is sometimes the best choice for the home gardener with very restricted space. Under this system all runners are snipped off as soon as they appear, so that you never have any daughter plants. All the vigor of the mother plant is thrown into berry production, and the result is the largest and finest berries of all systems.

If you decide upon this method, space your plants about twelve inches apart in rows two feet apart. Commercial growers usually space their plants twelve to fifteen inches apart in all directions, sometimes allowing even more space than this between rows to permit cultivation by machinery. The large berries produced by this means are sold to the "fancy trade." With good soil, moisture and care, you can grow berries just as fancy, but, remember, if a mother plant dies a vacant space will occur in the row.

Raised Strawberry Beds

Raised beds are nothing new and have been used for years because of the special advantage they provide for adequate drainage. However, ingenious gardeners constantly are coming up with new ideas.

One way of making a Strawberry bed is out of old railroad ties. How high you will build this depends on what you find comfortable if you want to sit down and pick. The large size of the ties makes this very convenient. Before filling your "box" with earth, however, be sure to put in a covering of heavy plastic, or use some other means of keeping the soil from coming into direct contact with the ties, as the preservative used in them may be detrimental to your plants.

If you find bending hard on your back, build a bed you can reach! Make boxes of redwood lumber two feet wide, ten inches deep, and about four feet long and set them on legs at a convenient height and

Hill system—all runners are kept clipped off. Space plants 12 inches apart.

Hedgerow system—each mother plant is allowed to set two daughters within the row. Place plants about 2 feet apart in rows 3 feet apart.

Double hedgerow—each mother plant is allowed to form four runners, or daughter plants.

place against a southern wall. Make as many of these as your space will allow and plant them to Strawberries. Then plant, weed, water and harvest, all at an easy reach!

Growing Strawberries From Seed

If you can't find a wild Strawberry patch, a wonderful way to get some of the wild Strawberry flavor into your jams and jellies is to grow the Alpine or Hautbois Strawberries.

These actually are distinct varieties of the common Strawberry, and are species native to Europe. The Alpine types come from *Fragaria vesca*, variety semperflorens; the Hautbois Strawberry from *F. moschata*. The berries have an exceedingly delicious flavor, are small but are borne over a long period during summer and fall.

Both types are propagated by seeds sown in the spring. Plants that are started early will bear fruit the same season. Start them indoors and transplant the young plants to flats of good, porous soil. Later they may be used to edge or border paths, walks or driveways. Most will do well planted twelve inches apart, but Baron Solemacher needs eighteen inches of room between plants. The Baron is a vigorous grower and does not form runners.

The Alpine variety *cresta* forms runners. *Fragaria vesca*, variety variegata, is very dressy, the handsome leaves being variegated with white. It also bears small, edible fruit.

Alpine Pineapple Crush (Park) bears large, 1-inch, cream-yellow berries of exceptional pineapple-strawberry flavor, borne on bushy, runnerless plants. They are distinguished by a marvelous aroma and are the earliest to bear of all the Alpines—as well as fruiting over a long season. Will produce the first year if sown in January.

Alpine Strawberry Mignonette (Park) is considered by many to be the best red Alpine Strawberry. It is also exceptionally prolific. The 1-inch fruits are delicious. Use it as an attractive edging or groundcover plant.

Sweetheart (Park), while not an Alpine, is a "first" for growing "real" Strawberries from seed: the seeds may be planted and Strawberries eaten the first year. These are big, everbearing Strawberries with real Strawberry flavor, tender and juicy yet firm, with a sweet delicious taste and aroma. It takes just 120 days from sowing to first fruit.

Strawberry Barrels and Jars

Like the recipe for frying fish, you must first catch a barrel. This may be a pickle or vinegar barrel or even a nail keg. If your barrel seems dry, soak it in water for several days and it will shape up again. If you suspect the inside may contain an acid residue, clean it thoroughly by using a solution of soda water or some other neutralizer.

With a two-inch auger, bore three or four holes in the bottom of the barrel. Then, starting about ten inches from the bottom depending on the size of your barrel, cut evenly-spaced holes around the circumference. Space the holes ten to twelve inches apart completely around

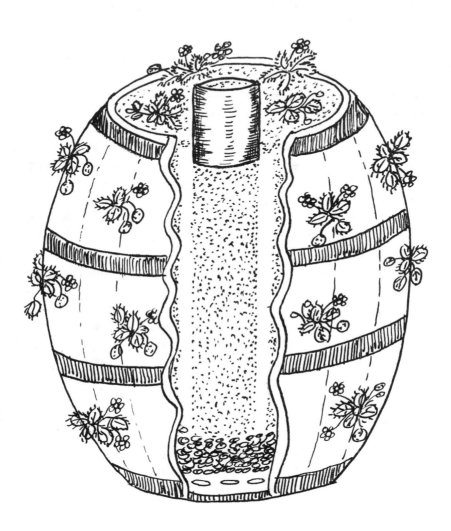

the barrel and about eight inches apart vertically as you move upward. Stagger each series so that every hole will be centered between the two holes just below. If you use a power saber or hole saw instead of the auger the job will be easier.

Whatever method you use for cutting, do not make the holes too large, or you will have trouble with soil washing out. Some people prefer to make the holes only one-and-one-half inches in diameter for this reason, but if the plants are properly set a two-inch hole will prevent crowding.

Select the spot carefully where you will place your barrel. Sunlight is important for successful growth, and the barrel must be given a half-turn once or twice a week. If it is too large to turn easily, mount it on furniture casters (three should do), an old wagon wheel, or even a clothesline turntable. If turning is not feasible, set plants only on the east, south and west sides of the barrel.

If you live in an especially hot, dry, windy climate the barrel should be protected from the hot summer sun in the afternoon. The east side of a building or in the shade of a tree would take care of this.

Fill the bottom of the barrel with six inches of gravel, small stones or broken flower pots. This will provide drainage for excess water.

In your garden cart or in a couple of bushel baskets, make up a rich soil mixture, using equal amounts of compost, well-decayed manure and sandy loam. If you cannot easily obtain these ingredients substitute sand, peat and dried manure purchased at a garden center. To this add liberal amounts of cottonseed meal. While the mixture should be rich, don't overdo it. Too much nitrogen will produce excessive leaf and runner growth at the expense of the berries. Cottonseed meal is particularly good because it releases its nitrogen slowly, making it available to the plants for sustained production.

A sand or gravel core is important, because it will permit water to move to the lower rows of plants. This core may be constructed easily by using a five or six-inch diameter tin can with both ends removed.

Place the can in the bottom of the barrel and fill with coarse sand or gravel. Pack the soil around the core up to the first level of the planting holes. The can is moved gradually up and refilled with sand as soil is packed around it.

The plants are set through the holes as the barrel is being filled.

Place them inside the barrel with some soil and carefully guide the leaves through the hole. Then, push the roots and soil up against the inside of the barrel. Work the soil well into the planting hole and pack in more to fill. Spread the roots inside the barrel as you do this, so that they are angled slightly upward. This will allow for settling of the soil.

As you pack the soil around the roots and into the hole be very careful not to bury the crown, as this can smother the plant. Continue adding layers of soil and building the gravel core until all the planting holes are filled. Set a few plants in the top, spacing them about eight inches apart. Leave the center (sand core) open for watering. Water thoroughly when you finish planting and often enough thereafter to keep the soil reasonably moist.

The variety of Strawberry that you will grow will depend on your own individual taste, but most people who make Strawberry barrels seem to prefer the everbearers, and Ozark Beauty and Ogallalah are the favorites.

Strawberries in barrels need winter protection in most climates either by bringing them into a building or by wrapping them with six inches of straw or other mulching material. Bales of old hay may be placed around the barrel. If you experience long dry periods during the winter, the berries should be watered, being careful to do this when temperatures are not too low. You may also wrap your barrel with burlap just as you do your tender shrubs.

Strawberry plants in barrels form runners just as other Strawberries do on flat land. When they start putting on, either pinch them off entirely or let them stream down the sides of the barrel. If you have lost plants from any of the holes use these to fill in with. You can root extra plants in milk cartons cut in half as previously described.

Strawberries grown in nail kegs or clay Strawberry jars may be grown in exactly the same way as in barrels. Strawberry jars are very decorative and may even be placed in a sunny window or put under grow lamps.

Strawberry Rings or Pyramids

This is one of the prettiest and most delightful ways to grow Strawberries and is my own personal favorite. The three-tier rings, the largest of which is six feet in diameter, form a terraced garden easy to care for in all respects. They come equipped with their own sprinkling system to which your garden hose is attached. We use a "Y" connector which enables us to water two rings at once and we also have equipped all our hose watering systems with Quick-Connect couplings that eliminate the need for threading, twisting and turning. You simply pull the band back to open and release to close.

Pyramids consist of three rings progressively smaller. We fill the first ring, packing it firmly with good soil and then allow it to settle for a week or two before adding the second. When this has settled we add the third. The short piece of hose to which the sprinkler system is attached must be placed when the first ring is filled in. Be sure to cover both ends with masking tape so dirt will not get into it. Adjust it carefully as you go along so it will be as nearly in the center of the third or top ring as possible.

These rings will accommodate fifty plants and each plant of the spring-bearing type may be expected to produce from one pint to a quart of berries in a growing season. How about that? Another advantage of rings is the ease with which they may be protected from birds. Frames and nets especially designed to cover them may be purchased.

The Everbearers

The Ozark Beauty Strawberry has stood the test of time and is ideal for growing in either a strawberry terrace or a patio tower. The strawberry terrace gives a maximum crop in minimum space. You can grow 50 Strawberry plants in just a 6-foot-square area. And the small space makes it easy to care for and pick. The Ogallala, another fine everbearer, is great for a strawberry tower that will accomodate up to 50 plants in as little as 4 square feet, and is movable if desired. These towers are constructed of rot-resistant wood with a water reservoir that gradually feeds and waters the plants. Store it for winter and use it again the following spring. There are even smaller ones now that take only 2 square feet of space.

Selva, another everbearer, new with Hastings, is a patented California variety believed to out-perform other everbearers under good conditions. It is much larger than Ozark Beauty, although not as tart. Runners are prolific. It is especially good for preserves and freezing. Grows well in zones 5–9.

Ft. Laramie, a new release from the USDA, is cold hardy, having survived -30°F., and is virtually disease-free. It has firm, large, sweet berries and high yields.

Quinault (Burgess) is a new release from Washington State and is one of the largest berries grown—often as big as teacups.

Growing Wild Strawberries

My husband, who is a "mighty hunter," likes to go to Colorado or Utah during deer season and we usually park our trailer somewhere in the mountains. While he hunts, I love to roam about gathering rocks, lichens, moss, wild rose hips and exploring for plants. On one such trek I came across a wide patch of wild Strawberries. There were so many that I didn't feel guilty about taking a few, so I dug them carefully, filled some discarded juice cans with leafmold and placed them therein, watering them carefully. They were in good condition when we arrived home later in the week.

In preparing my bed I tried to duplicate woodland conditions as nearly as possible, spreading my compost to a six inch depth, also incorporating eggshells and coffee grounds in the soil which I believe helps to keep down the depredations of cutworms. I placed the bed where the Strawberries could receive some shade, just as they did in their natural environment. I planted them twelve inches apart and mulched them lightly with leaves. Those hardy little plants never even knew they had been moved, and the following spring I was rewarded with a multitude of little white blossoms with bright yellow centers.

The berries, irregular in shape, were much larger than they would have been in the wild, but still were small compared to garden varieties. The flavor, however, is outstanding, and once you have tasted wild Strawberries I think you, too, will want to find a place for a little patch. You must be gentle in the picking and nip them off just below the calyx. This eliminates hulling later on. As with all Strawberries, they should be handled as little as possible and washed quickly under running water. Drop on a paper towel to remove excess moisture.

Wild Strawberries set runners just as domesticated varieties do, and if you wish you may increase your stock. Another advantage of the wild ones is that these little tough guys largely are untroubled by insects or diseases.

While you will not receive berries as large or as numerous as those from your garden varieties and the bearing season is short, they are very worthwhile. Mixed with other species they impart a special wild flavor unobtainable in any other way to Strawberry preserves—that touch of the unusual we all value and constantly seek.

SOME STRAWBERRY SECRETS

• If you want to stimulate runner growth, dig in some cottonseed meal around newly set plants or on a renovated bed.

• Excellent fertilizers for Strawberries are cottonseed meal, bone meal, and wood ashes. If a soil test indicates over-acidity, add agricultural lime.

• You can raise Strawberries on the same land year after year if you will keep it well supplied with organic material and the plants remain disease-free. The plants themselves should be renewed every two or three years.

• You can set new plants even in midsummer by using a bit of extra care. Using a large nail, punch a hole near the edge in the bottom of large juice cans. Set the cans as close to the newly set plants as possible. I like to sink the cans to a depth of two or three inches in the soft soil, which I have already rotary tilled before setting plants. Water plant well when set and mulch. Thereafter, unless weather is extremely hot and dry, water about once a week, filling the can to the brim each time.

• Always remember that one of the most necessary requirements for good Strawberry production is ample moisture and adequate drainage, coupled with good soil high in humus.

FROST CONTROL

Here in Oklahoma, my early blossoming Strawberries frequently are caught in a late freeze. If I have advance warning, I cover them with straw, burlap bags, old blankets, etc. If not, and the frost is unexpected, I sprinkle them with water early in the morning (when the temperature at plant level reaches 34°F) before the sun touches the frost. I continue to water until temperatures are above freezing in the morning and all the ice has melted. This often saves the blossoms. You can tell if frost has done its work by the centers: frequently, the white petals will appear unharmed, but the lovely gold centers where the berries are formed will turn black.

Here is another way to outwit Jack Frost. Place your Strawberry patch, particularly if you are growing the June-bearing type, at the highest point in your garden. Here it may escape harm, since frost tends to collect in low pockets.

GATHERING STRAWBERRIES IS FUN

Picking Strawberries probably is the most fun of all, and during the growing and bearing season I like to do this every day, sometimes even twice a day. Berries that show just slightly green in the morning often are completely rosy-cheeked by evening after the warm spring sun has touched them through the daytime hours. Strawberries, by virtue of the fact that they are so easily digestible, also are perishable and should never be allowed to melt on the vine by remaining too long.

When picking Strawberries I like to use a shallow bowl. Piled too heavily on top of each other tends to put too much pressure on the ones beneath, especially if the berries are very ripe.

If your planting is large enough to make a carrier convenient, make one out of scrap lumber. A shallow box, 11 inches by 16½ inches and about 3 inches deep will hold six quart cartons. For a handle use a couple of upright 1 x 2s, twelve inches in length. With small nails, attach one end to the box at the center. To the upright pieces at the top nail a 1 x 2 x 20.

Leave the caps on your berries, for they impart a special sweetness and should not be removed until you are ready to use them. Refrigerate berries for an hour or more, remove cap and wash in cold water to which you have added a few ice cubes. Drain on paper towels.

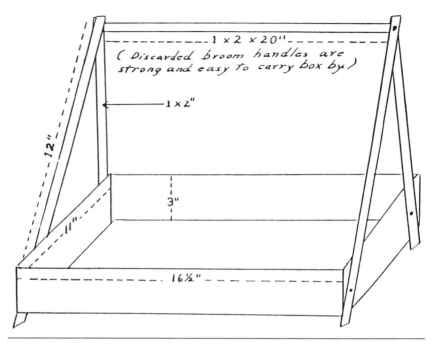

Carrier designed to hold 6 quart cartons.

STRAWBERRY VARIETIES

Lists of Strawberry varieties suitable for planting in different sections of the country change rapidly, since new ones are being developed and tested by Experiment Stations in many states. For the latest information you should call on your State Agricultural Experiment Station or talk to a strawberry-growing neighbor to find out what does best under local conditions. Many of the old standbys still are hard to beat.

For the **northern sections** the favorites seem to be Catskill, Midway, Midland, Sparkle, Robinson, Fairfax, Earlidawn, Howard 17 (Premier), Sunrise, Gala, Earlibel, Suwanee, Red Chief, Raritan (N.J. 857), Garnet (N.Y. 430), Surecrop, Redglow, Red Coat, Guardian, Jerseybelle, Vesper, Ozark Beauty (everbearing), and Geneva (everbearing).

For the region around **Washington D.C.**, on to the **Carolinas**, there are Blakemore, Klondike, Massey, Suwanee, Dixieland and Pocohontas.

For **Kentucky** and **Tennessee**, try Tennessee Beauty and Aroma.

In the **Gulf Coast states**, try Blakemore, Klonmore, Klondike and Missionary. These same varieties will do well in **Texas**.

For the **Pacific Northwest**, try Marshall, Brightmore, Corvallis, Northwest, Ozark Beauty (everbearing), Jumbo, Burgess Spring Giant, Burgess Hybrid No. 41 and American Sweetheart.

Varieties suitable for cultivation in **California** are Shasta, Donner, Tahoe, Lassen, and Sierra.

In **Illinois**, Blakemore, Fletcher, Midland, Midway, Pocohontas, Red Chief, Sparkle, Surecrop and Tennessee Beauty are favored.

Arkansas likes Albritton, Tennessee Beauty, Citation, Earlibel, Surecrop, Blakemore, Earlidawn, Sunride, Md. U.S. 2593 and Md. U.S. 2713

Florida has favorites also. These are Florida Ninety, which has an unusually long, conical shape, Tioga, Daybreak and Torrey.

For **Oklahoma**, the Oklahoma Experiment Station recommends Tennessee Shipper, Blakemore, and Pocohontas. I have grown all of these successfully. I am now getting my largest crops, however from Burgess Hybrid No. 41.

SOME STRAWBERRY RECIPES

Frozen Strawberries
Always select varieties of firm texture, well-ripened and of good red color. Flavor is important. Wash in iced water, removing hulls. Berries may be either crushed or sliced. Use ¾ cup sugar to each quart (1 ½ pounds) of berries, mixing throughly. Package, seal and freeze.

Strawberry Preserves

3 cups Strawberries   2 cups sugar

Select berries that would be slightly underripe for eating. Clean, wash and drain carefully to avoid crushing. You may cut berries or leave them whole.

To 1 cup of berries add 1 cup sugar. Heat gradually to boiling and boil 6 to 7 minutes. Add another cup of berries and another cup of sugar. Boil again for 6 to 7 minutes. Add last cup of berries and boil 7 minutes. Pour into clean, sterile jars and seal with paraffin. Yield is 1 ½ pints.

Strawberry Wine

Wine-making equipment for home use on a modest scale is now easily purchasable, and directions come with the equipment. If you would like to try your hand at making Strawberry wine, here is a very good recipe that will yield 12 quarts.

Measure 16 pounds of Strawberries. Crush berries. This will yield about 7 quarts of mash. Add 3 quarts of water and 4 ½ pounds of sugar. Add additional water to make 12 quarts. Add 4 yeast tablets—either Tokay or all-purpose yeast. Add 30 grams 80% lactic acid.

Strawberry Wine Jelly

This is a basic recipe that may be used for Strawberry wine or any other fruit wine. It's especially good when the cold winds blow, and it will bring a breath of spring to breakfast on a winter's day. This recipe may be doubled if you wish.

2 cups Strawberry wine	½ bottle liquid fruit pectin
3 cups (1 ¼ lbs.) sugar	(I use Certo)

Stir wine and sugar together until thoroughly blended in a glass saucepan, using a wooden spoon or paddle. Set over direct heat (medium) and stir constantly for about 3 minutes. Remove from heat, add fruit pectin and stir well. Pour at once into hot, sterilized glasses and cover with hot paraffin. This is especially attractive for gifts if poured into small brandy glasses. And it's perfect for serving with wild duck, goose or game.

4.

Raspberries
Come Next

aspberries, which ripen shortly after Strawberries, are very popular in every section where they can be grown. Moreover, unlike Strawberries, plantings well cared for will produce good crops for ten years or longer.

The fruit types available are red, black, purple and yellow, and many varieties come early, late, mid-season and as everbearers.

Red Raspberries include Fallred, everbearer; Latham, late; Newburgh, midseason; September, everbearer; Southland, everbearer; Sunrise, early and Taylor, late.

Black Raspberries are Allen, early; Blackhawk, late; Bristol, mid-season; Cumberland, mid-season: Dundee, mid-season; New Logan, early and Morrison, late.

Purple Raspberries are Amethyst, mid-season; Clyde, late; Purple Autumn, everbearer. Sodus also is an old favorite among the "Purples," which are a cross between blacks and reds. Royalty Purple (Stark Bros.) is a great new all-purpose rasp-berry. This very large, delicious introduction from New York ripens in late July in zone 6. For zones 4–8.

Yellow Raspberries are Amber, very late; and Fallgold, everbearer. I have grown Forever Amber (Burgess),a yellow member of the Black Raspberry family and find it particularly delicious. It is of medium size with sweet yellow-to-amber berries, which have a delicate Black Raspberry flavor and aroma.

Baba berry—a new one to try. It was developed by a personal friend of mine who found the original plants growing near her vacation cabin high in the mountains. She sent me some of her original plants and they grew well in southern Oklahoma. The everbearing Baba (Hastings) can take the heat, as well as being hardy all the way up to Michigan. The large, firm sweet red fruit is of excellent flavor. It is disease resistant, vigorous and productive. In zones 8 and 9 it bears a second crop.

Common name: Raspberry
Botanical name: *Rubus occidentalis* (black)
***R. idaeus* (red)**
Soil: Fertile, well-drained soil with plenty of humus, supplied either as a cover crop or well-rotted stable manure.
Nutrients: Too much fertilizer gives lush growth and few berries. Fertilize with well-rotted manure in the spring.
Water: Keep fairly moist but not wet. Do not let them dry out.
Spacing: 3 feet apart (red); 3 to 4 feet apart (black).
Sunlight: Full sun. In hot summer areas they take light shade.
When to plant: Purchase bare root in early spring or in containers in early summer. May be planted in late fall in milder areas of the country.
When to prune: After fruit is harvested in the summer, cut 2-year-old canes to the ground, selecting 5 or 6 healthy new canes for new growth (these are cut to within a few inches of their supporting wires).

Raspberries are native to North America, Asia and Europe. They were known to the ancient Greeks, who called Raspberries "Idae" because they grew on Mount Ida, in Crete. Linnaeus called the European Raspberry Rubus idaeus, "the red berry of Mount Ida."

Names of Raspberry varieties seem almost endless when you look through nursery catalogs, each one having its own good qualities and differing from the others in size, quality, color or bearing season. Again, as with Strawberries, I would urge a little detective work on your part, before you purchase. Find out what will grow the best in your area and select types best adapted to your section of the country. Your County Agent or State Experiment Station can help here also. One-year-old, No. 1 grade plants are best for establishing new plantings. Make every effort to secure virus-free plants.

HOW RASPBERRIES GROW AND BEAR

Moisture supply is the real key to success in growing Raspberries. If there is a choice of location, pick a spot where there will be enough moisture at all times.

Drainage, considered from the viewpoint of both air and water, also is important. Remember that cold air will settle to lower levels. Crops located on land higher than the surrounding fields will be less likely to suffer from frost injury. Too, free movement of air during the growing season will help to lower humidity. High humidity may create conditions conducive to the growth of fungus diseases.

Raspberries are best planted in early spring (late March or early April) in northern sections. Fall plantings will do well farther south if they can have time to become established before cold weather.

Plants never should be allowed to dry out. Prevent this by placing them in a bucket of water or dipping the roots in a thin clay mud.

Always have your planting location prepared well in advance, being sure to incorporate plenty of humus in the soil, getting this in either as a cover crop or well-rotted stable manure. If manure is unobtainable, sow rye or buckwheat and plow it under when the growth is thick. If you have a rotary tiller, go over the entire area with it before planting, thoroughly mixing the contents of the soil. Then let it settle for a week or two.

Do not plant the Raspberries if the soil is excessively wet from spring rains. Store the plants in a cool place instead and wait until things dry out. Then your soil will not pack.

This is very important because of the **leader buds**. When your Raspberries arrive from the nursery you will note several small, pale colored, yellow-green growths on the stems just above the roots. They may not seem to be very important, but once set in the ground they soon will push their way upward and become little shoots. Packed soil will slow them down or make it impossible for them to come through. These shoots in Raspberry language are called **canes**. They must hurry and grow very fast, for their lifespan lasts only for a little more than a year.

When ready to plant, carefully spread the roots in the planting hole and firm the soil over them. Water to prevent air pockets.

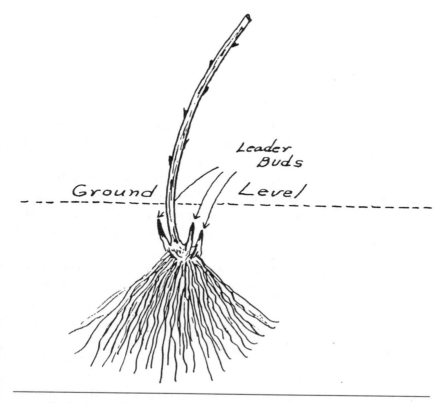

Newly planted Raspberry plant showing position of leader buds just above roots. Set Red Raspberries 2 to 3 inches deeper than they were in the nursery, set Black and Purple about one inch deeper.

Set red Raspberries 2 to 3 inches deeper than they were in the nursery, and set black and purple Raspberries about 1 inch deeper.

Red Raspberry plants should be cut back to 8 or 12 inches after planting. The canes, or "handles," of black and purple Raspberries should be cut off at ground level, removed from the planting and burned.

If there are any wild brambles growing around or near the new planting, they should be dug up and destroyed to prevent the possibility of their carrying disease.

Depending on how much space you have, you can plant Raspberries from three to four feet apart. Four feet is better if you have the room, since some varieties make a very vigorous growth. If you plan on having more than one row, make the space between rows six to eight feet apart.

MULCHING

Generally Raspberries should be cultivated during the early part of their first summer. Then in late summer, after your plants are established, they may be mulched. Mulched Raspberries grow better, produce more and have larger berries. The mulching materials may be straw, crushed corncobs, leaves, grass clippings, or sawdust. If you use sawdust and it is not thoroughly decomposed, add some natural nitrogen such as cottonseed or soybean meal to prevent nitrogen tie-up.

Mulch four to eight inches deep is needed to keep down weed growth, and you should continue adding mulch material each year as it is needed.

TRAINING AND PRUNING RASPBERRIES

Understanding how Raspberries grow and bear fruit helps us to understand why they should be trained and pruned by certain methods.

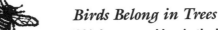

Birds Belong in Trees

If birds are a problem in the berry patch, a good way to deal with the berry stealers is to plant wild fruits on your place, as the birds like these better than your tame fruits. Usually, birds eat Raspberries, Strawberries, and other sweet fruits only when more favored fruit is not available—they tend to like bitter, sour fruit. So get some chokecherries, bittersweet, highbush cranberries, and similar wild fruits established. It will take a lot of pressure off your berry patch.

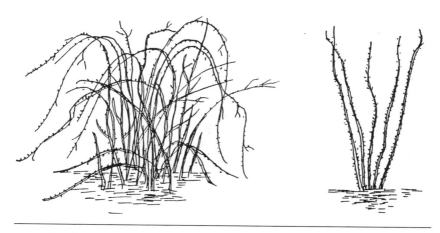

Raspberry plants left unpruned will become a mass of brambles. Unpruned plant at left is shown at right after being properly cut back. Head canes of Red Raspberries back to about 30 inches.

They bear their fruit on biennial canes, but the roots and crowns are perennial.

All brambles send up new shoots or canes during each growing season from the crown of the plant. Red and yellow Raspberries develop new shoots from both the crown and roots.

These new canes, whether from crown or roots, will grow vigorously during the summer, initiate flower buds in the fall, overwinter and then bear the following season. This fruit is borne on the leafy shoots which arise from lateral (side) buds on one-year-old canes. Once having borne their crop, the canes start a gradual process of drying up and begin to die back shortly after the fruit is harvested. They should be cut and burned. New shoots soon will begin developing to repeat the cycle and provide fruiting canes.

PRUNING EVERBEARERS

While cutting the canes after they fruit and die back is the approved method for most Raspberries, the everbearers have a different lifestyle. They bear fruit twice on the same cane. The new shoots bear a crop at the tips in the fall and bear again the next season further down on the canes.

For this reason, the fruit canes of the everbearers should not be pruned after bearing the fall crop, since this would remove the fruiting wood for the spring crop.

For those who prefer one large crop to two smaller ones there is still another way. Everbearing varieties will produce abundant fruit on **primocanes** (canes of the current season's growth), and thus it is possible to grow them only for the fall crop. This is accomplished by mowing all of the canes to within two or three inches of the ground in the spring before growth starts. In the fall only the abundant crop on the primocanes is harvested. There are several advantages in doing this: first it eliminates all labor of hand pruning, and secondly winter-injured canes present no problems. Fungus diseases are held to a minimum as well.

GROWING RED RASPBERRIES

Because of their need for moisture, Raspberries as a whole are better adapted to the cooler northern climates than the warmer southern sections. Most red Raspberries are grown in the northern states, but in the South, at higher elevations, it is possible to produce them. At present I am growing Mammoth Red Thornless and experiencing no difficulty. This variety not only will do well here in the Southwest but will withstand the bitter winters of Minnesota. Latham is another good red Raspberry for the North.

Varieties of Raspberries differ somewhat in size and color and season of fruiting, but all require abundant moisture. Not even Strawberries are more demanding in this respect.

Because of this, a successful planting must have ample amounts of organic matter incorporated in the soil. At the same time you must provide for drainage, for no Raspberry will tolerate wet feet.

Spring planting in the North or fall planting in the South will both give good results if the other requirements are fully cared for. Simply set them in the soil a few inches deeper than they grew in the nursery row, firm the soil and water if necessary.

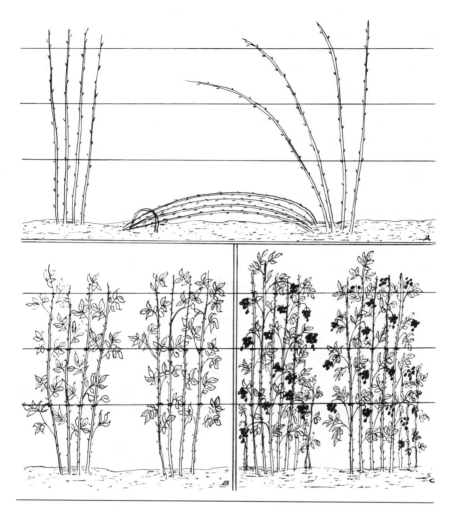

Growing Raspberries in the North

A. In the fall, after the first killing frost, prune long Raspberry canes back to 6 feet. Bend canes down and lay them flat on the ground, parallel to the row, securing them so they will not pull back up again.

B. Straighten canes after uncovering and tie to the supporting wires. Cut them back if they are longer than 6 feet.

C. In early summer new canes will start from base of plant. Again choose four healthy sturdy cares and allow them to develop. Try to choose cares about 6 to 8 inches apart.

In training the Reds you either may keep all suckers pulled (which will limit your number of plants to just what you started out with), or you can let some of the suckers grow within the row to form a hedge. For a hedge, set your plants three feet apart; for individual bushes set them four feet. For either, if you plant in parallel rows, space the rows five feet apart.

The soil should be kept well cultivated the first few months after planting. Then, about the middle of the summer after the canes for next year's growth are well up, start applying mulch around them. This will help to conserve moisture and keep down weed growth.

The following spring all of last year's dead canes should be cut out and removed. The new ones should be headed back to about thirty inches. They will start putting out lateral branches quickly and these will bear fruit in midsummer. Leave five to eight fruiting canes per hill for mature plants, and at this time also remove any weak canes.

The first crop will be rather small, but who could resist the fun of gathering it? It will be a preview of things to come, for your red Raspberries will really get going when they come into the third year of production.

Red Raspberries properly pruned usually are able to stand well without support, but if they need help here is what to do: drive a stake beside each plant and tie the canes loosely to it. Do not bunch them up tightly or they will not have good air circulation. If you are growing your Reds hedge-style, set posts at the end of each row, stretch a wire between them and tie the canes to the wire.

Once your plants are well established, pruning and thinning will be about all that is necessary to do. And it couldn't be more simple. All it involves is cutting out all the old canes as soon as they have borne fruit. There is no point in keeping them, for they will never bear again. They should be cut close to the crown and burned. Thinning the canes will be necessary each year, again leaving five to eight. Try to space them at least six inches apart.

To keep your Raspberries within bounds you will have to control their enthusiasm for producing unwanted suckers. With the hill system, these should be pulled up from the very start. If you are letting your Raspberries establish themselves as a hedge, let the suckers develop in the row about ten inches apart. Pull up any that appear at the sides. Once the hedge is established keep all new ones pulled.

And when I say "pull up," I mean PULL UP. Cutting them off just makes them more eager to start growing again. And never forget that if you cut a root of a red Raspberry you will have a sucker springing up right there. So it's a good idea to depend upon mulching instead of cultivation to keep down the weeds.

If Red Raspberries are adaptable to your climate you can propagate indefinitely from the first variety you plant. Just dig up a healthy sucker and plant it in a new location.

GROWING BLACK RASPBERRIES

My first experience with growing black Raspberries was with Morrison, and at first I was delighted, then appalled, by their eagerness to propagate themselves. They are constantly trying for parenthood, arching over their canes and burying the tips in the ground. Every time they do this, especially in good soil, the tips take root and produce new plants. These tip plants, if allowed to form unchecked, soon will fence you in, and your Raspberry patch will become a veritable jungle. There is no need to let this happen, for they can be properly restrained.

Blackcaps, once well established, will yield about a quart of berries per plant each season. How many plants you will need depends on the space at your disposal and how much you enjoy the fruits.

Those who know best about such things advise that if you grow both Red Raspberries and Black Raspberries you should put a considerable distance between the two types. The reason for this is that the reds sometimes carry a disease that does little or no harm to them but may prove near fatal to the blacks.

In *Hints for the Vegetable Garden*, compiled by the Men's Garden Clubs of America, tests showed that sawdust and wood-chip mulches increased raspberry production by as much as 50 percent. Apply 3 to 4 inches of this mulch at the base of the plants.

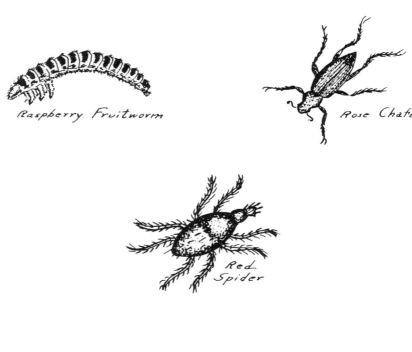

Pests Injurious to Small Fruits

Raspberry Fruitworm—several species. Adults: Yellow to brown beetles; ¼ inch long. Larvae: Brown and white; up to ⅛ inch long. Damage: Adults make long, narrow slits in blossom buds and newly formed leaves; larvae feed in berries. Rotonone dust or spray is effective when applied 3 times at 10-day intervals, starting 7 days after the first blossoms appear.

Rose Chafer: Gray or fawn-colored beetle; reddish-brown head; long-legged and slender; ½ inch long. If not present in large numbers, remove by hand. Damage: feeds on foliage, buds, flowers and fruits of Blackberry and Raspberry.

Red Spider—several species. Adults and young: tiny (barely visible to the naked eye); red or greenish red. Found on undersides of leaves. Damage: yellow specks and fine webs on leaves; plants and fruits are stunted. Control: malathion. Partial control may be achieved by applying a dust containing 25 to 30 percent of sulfur or by applying a spray containing sulfur.

White Grubs—several species. White or light yellow; hard brown heads; curved; ½ inch to 1 ½ inches long when full grown. White grubs live in soil and are larvae of May beetles. They require 3 years to mature. Damage: Larvae are injurious to strawberry roots. Land that has been in sod often contains white grubs. Plant such land to something else for a year or two first before planting strawberries on it.

Grasshopper—several species. Adults and nymphs: brown, gray, black or yellow; strong hind legs, up to 2 inches long. Most grasshoppers are strong flyers. Damage: When present in great numbers will destroy leaves of berry plants. Apply a dust or spray containing malathion.

Black Raspberries are less hardy than the reds, more easily harmed by cold weather and, for this reason, should be planted in the spring.

Blacks should not be planted quite as deeply as the reds, an inch or so deeper than they stood in the nursery row is sufficient. But give them more space: five feet apart within the rows and the rows also five and six feet apart. Cultivate the soil and follow a mulching program just as you did for the reds.

Pruning should begin with the blackcaps as soon as the new shoots are eighteen to twenty inches tall. At this height the top is pinched off to make the canes branch. This makes them sturdier and easier to manage. If this is not done, the canes will become long and sprawling. This operation, called tipping, not only will prevent the bush from growing taller, (resulting in a small, compact bush), but it also will keep your blackcaps from getting out of control by forming tip plants. And it makes them put out laterals which will bear the following year.

By late summer or early fall the laterals will be several feet long, and a number of fruit buds will have developed. During the winter both the canes and the laterals will "sleep" in what is called dormant rest.

In the spring, while the canes are still asleep, the laterals should be cut back to five or six fruit buds. You have a choice: the more fruit buds you leave the more berries you will have; the fewer the buds the larger the fruit.

Blackcaps, like the reds, have a way of repeating their pattern. As the old laterals get ready to bear, you will notice new shoots springing from the leader buds at the bases of your plants. Let these grow to the same height as before and tip them. These, in turn, will produce laterals which will bear a crop the following year.

After your first crop of fruit has been gathered, cut out the old canes and burn them. Leave the new canes as before until the next spring, and then shorten them.

As your plants become well established, you will have more new canes than you may want, and these will need to be thinned out. There is no hard and fast rule as to how many, but experience has shown that four to six of the largest canes will bear the best crop.

If you want to increase your stock, let some of the canes grow without tipping. If they act contrary and do not bend down naturally and form new plants, bend them down and anchor them; then cover the tip with earth. By the following spring you will have new plants that may be cut loose and reset in a new location.

HARVESTING

The very delicacy of the Raspberry that makes it so delicious and desirable also makes it very perishable. Pick the fruit every two or three days when fully ripe. The reds are best when they are a deep garnet and begin to push away from the stem.

To obtain the finest flavor, Raspberries should be picked in the late afternoon. Pick in small baskets and do not press them down. Pick only the best, for often there is a great deal of difference in the quality of berries growing on the same plant.

Remove your berries to the shade or a cool storage place as soon as possible. Raspberries, if washing is necessary, should be placed in very cold water and removed quickly. Serve at the table well chilled, and in a shallow dish, so they will not be crushed. I like to use a large saucer or shallow bowl so they can be seen and admired.

RASPBERRY PROBLEMS

Raspberries are lucky in having few pest and disease problems. **Mosaic** and **leaf curl** largely can be avoided by buying only clean stock and keeping reds and blacks widely separated. If space is limited, it is best to plant only one kind. If disease does occur, burn all infected plants. **Red spider mites** often can be kept under control by washing the plants off with a stream of water, but if they appear in large numbers, dust with malathion.

I try to keep a close watch on my plants and catch everything early. Sawdust will indicate a **raspberry cane borer**, the grub of the long-horned beetle. I cut until I find it and then destroy the grub. **Root borers** will cause canes to wilt, and these should be dug out and the canes burned. **Cane maggots** sometimes girdle new shoots causing the tips to wilt. Cut below the girdle for several inches, and if there are eggs in the pith burn them. You need not remove the cane. Let it bear as usual after this operation.

Japanese beetles are troublesome in some sections. Ask your County Agent the best means of control, or try Milky Spore Disease, now obtainable from many nurseries and garden centers.

In general, cleanliness is your best guard against raspberry diseases. Any wilted, rusted, moldy or discolored canes are under suspicion. Play safe; cut them out and burn them. Keep the weeds down, do not allow trash to accumulate, and remember that good air circulation is important.

RASPBERRY RECIPES

For the following recipes you may use any of the Raspberries: red, black, purple or yellow.

Raspberry Jam
From the Ball Blue Book

2 quarts raspberries, washed and dried
⅓ cup water
1 tablespoon lemon juice
1 tablespoon lemon peel

1 package powdered pectin
6 cups sugar

Combine raspberries, water, lemon juice, lemon peel and pectin in a large saucepot. Bring to a rolling boil over high heat, stirring frequently. Add sugar, return to a rolling boil. Boil hard 1 minute, stirring constantly. Pour hot into hot jars, leaving 1/3-inch head space. Adjust caps. Process 10 minutes in boiling water bath. Yield about 5 half pints.

Raspberry Ice Cream

1 cup condensed milk
¼ cup water
2 ½ cups fresh Raspberries, crushed
2 tablespoons lemon juice
½ cup heavy cream

Mix condensed milk and water. Add Raspberries and lemon juice. Chill. Whip cream until thick enough to hold a soft peak. Fold into chilled mixture and pour into freezing tray of refrigerator. Freeze lightly. When about half frozen beat until smooth but not melted and freeze again until firm.

Fresh Raspberry Pie

1 cup sugar
2 tablespoons cornstarch
⅛ teaspoon salt
3 cups fresh Raspberries
1 recipe plain pastry, or 9-inch pie shell
1 tablespoon margarine

Mix together sugar, cornstarch and salt. Add to berries. Line pie pan with pastry and add filling. Dot with margarine and cover with top crust. Pierce crust with fork. Bake in very hot oven (450°F) for 10 minutes. Reduce temperature to moderate (350°F) and bake 30 minutes longer.

Frozen Raspberries

Some happy year you will find you have all the fresh Raspberries you want to eat and enough left over to freeze for special treats. Here's the way to do this:

Select firm, fully-ripe berries of bright color. Wash quickly in ice water, sort and drain thoroughly. An unsweetened pack is satisfactory and berries may be sweetened with sugar or honey when served.

For a sugar pack, use ¾ cup sugar with each quart (1 ⅓ pounds) of berries.

For a syrup pack, cover berries with cold 40% syrup. Make this syrup with 3 cups sugar and 4 cups water. Dissolve the required amount of sugar in cold water, stirring solution occasionally to dissolve the sugar. If you do not need all the syrup at once it may be stored in your refrigerator for a day or two.

Blackberry blossoms have white petals tinged with pink. The fruit is attached near the stalk in a cylindrically shaped growth that comes off when the fruit is picked. Pick blackberries when they are black, before they grow too soft.

5.

Blackberries Are Beautiful

ears ago when I was a new bride, my young husband and I had our first real fight over a Blackberry bush. A volunteer plant had sprung up in our fence line and he ruthlessly cut it down.

To him that Blackberry was exceedingly unwelcome, the possible forerunner of a thorny thicket—moreover it was in the wrong place. To me, every berry, regardless of kind, size or color, was precious—a possible source of jams and jellies.

We finally patched things up by effecting a compromise. He agreed to dig up the root and put it in a more suitable location and I agreed to stop yelling.

Blackberries are one of Oklahoma's best food products—we are second only to Texas in production—and we take a back seat to no one in size and quality of berries.

Probably many of you picked Blackberries in your childhood and remember an enchanted wild briar patch where you gorged on the sweet fruit, staining your face and hands a deep purple, and getting all scratched up from the thorns at the same time. For Blackberries do grow wild in many places, and once they could be picked at will along roadsides. But no longer is it safe to do so because of weed killers being so liberally applied.

For those who feel nostalgia for the good old days, the Cascade Blackberry most nearly approaches the wild ones in flavor and also has the added appeal of being much, much larger. And they still have all the thorns you remember. You can get just as gloriously scratched up as you ever did.

This variety is a strong and sturdy grower, and will take over if allowed to have its own sweet way. Don't let it. Keep Cascade high up on a trellis, for any vine tip that trails on the ground will take root quickly.

For those who like their Blackberries without pain, Thornfree offers a way out. The fruits are medium size, firm and glossy and the plants should be set three feet apart.

For the northern states, DeSoto is an everbearng Blackberry hard to top. A vigorous grower, it starts to bear during the regular midsummer Blackberry season and continues until fall. The berries are large and very sweet.

At the present time the general favorite for all sections is Darrow, one of the newer and better varieties. This one also has a lot of the good, old-fashioned flavor, and it is good for just about any purpose, for eating out of hand, for jams, jellies, freezing or making wine.

Just one word of caution—if you do find a wild patch way out in the country and decide to pick it, do so but don't crowd your luck. All kinds of wild things love Blackberries too, and these include bears, snakes and, of all things, spiders.

Long ago I wondered what manner of creature was stealing my berries, so one morning when the pre-dawn light was just barely enough to see by I went out to my patch. I found several tarantulas, the big black, hairy spiders of the Southwest, gleefully harvesting. They would cut off a berry at a time and roll it into their holes in the ground. Tarantulas look ferocious but are relatively harmless, although some snakes are something else again.

While I still venture out and pick the wild ones occasionally, I can do my harvesting in greater comfort at home, where I grow the best domesticated varieties.

It won't take many plants to give you all the fruit you will want, since each Blackberry bush can be counted on to yield about a quart and a half of berries in a season.

Blackberry culture, as I said before, is almost too easy in sections where they are well adapted. They are one of the "cane" fruits, and this means, just as with Raspberries, that the canes, or stems, will bear fruit the year after they sprout. These canes then die and new ones spring up to replace them.

An established patch may bear for ten to fifteen years, even longer if well cared for. The bushes propagate themselves by sending up suckers from their roots. In this respect they are like Raspberries, but unlike Raspberries, which grow best in cool, moist, northern regions, Blackberries prefer milder climates and do best in the South. More shallow-rooted than many other berries, they must have ample moisture along with good drainage. Protection from drying winds is highly desirable.

If you live in the North, however, and want to grow Blackberries, don't give up, for a great deal of work has been done toward producing varieties that will also grow in colder climates.

Common Blackberry varieties are classified as hardy, semi-hardy and tender. Choose according to the part of the country where you live. Even if you have to pet your plants along a little and give them some winter protection, it will be well worth the time and trouble.

Common name: Blackberry

Botanical name: *Rubus ulmifolius* (thornless)

Soil: Grows best where loam is deep, mellow, and damp.

Nutrients: Well-rotted manure.

Water: Fairly moist, but not over-wet. Do not let them dry out.

Spacing: 3 to 5 feet apart, in rows 6 feet apart.

Sunlight: Full.

When to plant: Early spring.

When to prune: After fruiting. Running or trailing types must be trellised to prevent thickets. Thirty inches is considered best height for heading shoots back.

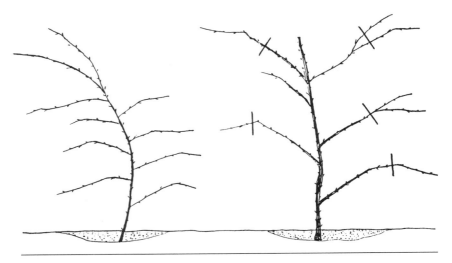

Pruning Blackberries makes the difference. Note sturdy plant on the right. This was pruned during the summer and will be pruned the following spring where the black marks indicate.

Having decided upon your particular varieties, order your stock so it will arrive in ample time for early spring planting if you live in the North. Fall or early winter is the best time to plant if you live in the South.

PLANTING AND CULTIVATION

You can increase the productive life of your Blackberries if you have your soil well prepared in advance. Depending on how good (or how bad) your soil is, incorporate goodly amounts of compost in it, mixing well. I do not advise the use of chemical fertilizer with any berries, and the over-use of it can result in the production of many canes but little fruit. Compost will improve the texture of the soil as well as supplying needed nutrients.

When your plants arrive, trim off any broken or over-long roots and cut the tops back to about six inches. Dig your hole, spread the roots out fanwise, insert, and firm the soil well with your foot. Fill in soil around the plant so it will stand just about as deep as it grew in the nursery. Blackberries are not as fussy as Strawberries, and an inch or so either way will not matter. Water to prevent air pockets about the roots.

If you plan to plant in rows you should leave three to five feet between the plants in the row, depending on variety. Blackberries are vigorous growers and take lots of room, so don't crowd them. At least six feet between rows is considered a good distance.

Cultivation is the order of the day during the first spring and summer. Do this often—at least once every two weeks or so. Do not go too deep; remember those shallow roots. Like the Raspberries, these plants, too, will sprout a sucker whenever a root is cut. Toward midsummer start mulching, putting on several inches of whatever material is available—hay, grass clippings or cottonseed hulls. This will keep weeds down and help to conserve moisture. If this is properly done your labors will be lightened—no small thing in my books, for I'm no work horse. I always do things the easy way if possible.

As soon as the young shoots get up to around two or three feet, snap off their tips. This will cause laterals to put on, and these will bear fruit the following year. Thirty inches is considered the best height for heading shoots back, causing the bush to grow strong and compact.

The following spring the laterals should be shortened to about a

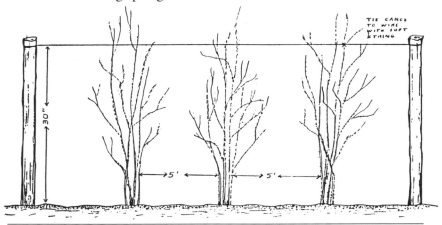

Erect Blackberry plants can be grown without support, but many of the canes may be broken during cultivation and picking. You can make a simple trellis by stretching wire between posts set 15 to 20 feet apart in the row. Use a single wire attached to the post about 30 inches from the ground. Tie the canes to the wires with soft string where the canes cross the wire. Avoid tying the canes in bundles.

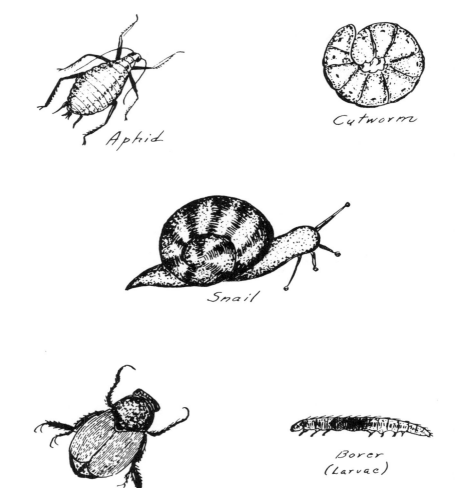

Aphid

Cutworm

Snail

Japanese Beetle

Borer (Larvae)

More Pests

Aphid: Adult and young: very tiny. Green to greenish brown. Soft-bodied. Covered with a fine, whitish wax. Aphids cluster on leaves. Damage: curled and distorted leaves. Dust or spray with malathion.

Cutworm—many species. Cutworms are dull gray, brown or black and may be striped or spotted. They are stout, soft-bodied and smooth, and up to 1 ¼ inches long. They curl up tightly when disturbed. Damage: cut-off plants, at or below soil surface. Some cutworms feed on leaves, buds or fruits; others feed on the underground portions of plants. You can prevent cutworm injury to many plants without using an insecticide by placing a stiff, 3-inch cardboard collar around the stems; allow this to extend about 1 inch into the soil and protrude 2 inches above the soil. If used for Strawberries clear the stem by about ½ inch.

Snail: Gray or black in color with rounded shells. They and a related species, Slugs (which have no shells) are often a great nuisance, especially in very wet seasons. Putting salt on the soft-bodied species dissolves them. Both snails and slugs can be trapped in beer cans or in saucers containing a small amount of beer.

Japanese Beetles: Adult: Shiny metallic green; oval; coppery brown outer wings; about ½ inch long and ¼ inch wide. Larva: White body; brown head, up to 1 inch long when fully grown. Damage: Adults may attack foliage of Raspberry, Blackberry, etc. Dust or spray with malathion. Or obtain Milky Spore Disease, which is sold by nurseries or garden centers for control of Japanese Beetle grubs.

Red-Necked Cane Borer: Adult: Dark bronze or black beetle, shiny, copper-red neck, slender, about 1 inch long. Larva: White; flat head, slender; up to ¾ inch long. Damage: Adults eat margins of leaves; larvae tunnel canes, causing spindle-shaped swellings on surfaces. Cut off infested canes well below the points of injury and destroy them.

---

foot and a half. As soon as new canes become evident, select four or five of the best and strongest, and remove the others by "rubbing," which means cutting them off at ground level. When the new canes are about thirty inches tall, tip them as you did the others the year before, thus preparing them to bear next year's fruit. You may need to go over the patch several times in order to head back all the canes to a proper height.

Expect a small crop in midsummer of the second year from the old canes (those that grew the first year). These should be cut off close to the ground and burned as soon as they have fruited, for they will never bear again. Getting them out quickly will improve air circulation and make room for the new shoots.

When the spring of the third year arrives, the laterals of the second year's growth should be cut back to one and one half feet. Thereafter, follow this same system. Select your best new canes, tip them back and shorten the laterals.

Blackberries usually are strong enough to stand alone, but some varieties may need support. If this is necessary tie them loosely to a stake. If you are growing them in rows the canes may be tied to wires stretched between end posts.

TROUBLES

Blackberry pests and diseases are likely to be the same as those affecting Raspberries, but are less likely to occur. Blackberries are just naturally very strong and vigorous.

There is, however, a problem that occasionally occurs and for many may be hard to identify. This is **sterility**. The healthy-looking, normal-appearing Blackberry plants may flower profusely and then fail to produce fruit. It may be a complete or partial failure. Sometimes this is evidenced by misshapen berries: some are almost normal in appearance, while others have only a single drupelet develop (drupelet are the individual nodes that make up the berry).

There may be several causes for this—hereditary abnormalities, insect damage, fungus infection, virus infection or even a combination of several factors. Gene or chromosome combinations that do not permit effective self-pollination may result in poor fruit set. Though the plants and flowers appear normal, the pollen produced may not effect fertilization of the ovules, which is necessary for normal fruit development. Sometimes the fruit malformations are caused by insect damage, but extensive damage from insects seldom occurs. Some varieties, also, are unfruitful if planted alone.

Anthracnose is the most common fungus infection that may affect fruit development, but it seldom is severe on erect Blackberries. Its evidence is the immature drupelets which, instead of ripening

Tame Those Brambles!
Training and pruning Blackberries and other bramble fruits need not leave you scratched and bleeding. Wear bee gloves, leather gloves with heavy cloth uppers that reach almost to you shoulders. Makes the job practically painless. For bee gloves write to A.I. Root Company, Medina, Ohio 44258-0706, or American Bee Supply, Inc., Route 7, Lebanon, Tennessee 37088-0555.

normally, become small, brown, dry and woody. Remove any suspicious-looking canes at once and all fruiting canes as soon as they have been harvested.

In Blackberries sterility also can be a symptom of a virus infection that affects the whole plant. Such plants will produce new canes that are, seemingly, more vigorous and with rounder and glossier leaflets than normal. It is indicated also by premature fall reddening of the leaves, which become a brilliant scarlet. Even the flowers on these plants appear to be normal and healthy, but only a few drupelets will develop in each receptacle. The fruit bud production for the following year also is reduced.

This virus will spread, but how it spreads (other than by root suckers) still is unknown. If your planting becomes infected do this:

• Remove plants that fail to set fruit, also grubbing out as much of the root system as possible, and burn immediately.

• Never use root suckers to propagate plants from any Blackberry fields where the sterility virus has been detected.

• Be sure to remove all old canes soon after harvest.

• Dig up and burn any neglected plantings or wild bramble patches found near new plantings.

• Purchase your new plants only from nurserymen who will certify their plants to be produced from fruitful stock.

BLACKBERRY RECIPES

Pinch any Blackberry lover and ask him (or her) to name his favorite dessert and he will immediately say Blackberry Cobbler. Here is one of the best recipes for it that I have ever found:

Blackberry Cobbler

2 ½ cups fresh Blackberries
4 tablespoons all-purpose flour
¼ cup sugar (more or less depending on ripeness of berries)
½ recipe for shortcake
½ teaspoon cinnamon

Mix flour, sugar and cinnamon. Blend with fruit. Pour into oiled casserole. Roll shortcake dough ½ inch thick. Fit over top of fruit. Prick with fork. Bake 25–30 minutes in preheated oven at 400°F. Serve hot. It's great with vanilla ice cream.

Shortcake

This basic recipe also is delightful with Strawberries. Have ready a 10 x 14 inch cookie sheet if used for shortcake. Preheat oven to 425°F.

2 cups all-purpose flour
    (sift before measuring)
4 teaspoons baking powder
½ teaspoon salt
2 tablespoons sugar
½ cup cooking oil
¾ cup sweet milk
2 tablespoons butter or margarine

Blackberry Jungle

Sometimes you can have too much of a good thing! Here in southern Oklahoma where Blackberries grow wild, they sometimes completely take over a wide area.

Cutting them down to size is not an easy task. If done by hand they should be cut out with a brush ax. If new shoots start to grow—and they probably will—they, too, should be cut down. Eventually the roots will die.

Moon Signers believe the best times to destroy "noxious growths" is when both Sun and Moon are in barren signs, and Moon decreasing; Fourth Quarter preferred. Barren signs are Leo, Aries, Virgo, Aquarius, Gemini, and Sagittarius (listed according to the degree of their nonproductive or barren qualities).

Sift dry ingredients together and work in oil until mixture is a crumbly mass. Add milk. When all ingredients are well blended, toss on floured kneading board. Knead gently for a minute or two, then pat to ½ inch thickness. Bake either as shortcake or use ½ recipe as directed for cobbler. Split and spread with butter.

Blackberry Jelly

Blackberries—half fully ripe and half underripe
Water
Sugar

Wash berries and place in kettle. Add 1 cup water for each 3 quarts of berries. Heat slowly to boiling and boil about 5 minutes. Strain through jelly bag. Add ¾ cup sugar for each cup of juice. Cook in small batches, only 3 to 4 cups of juice at a time. Boil juice 3 minutes, add sugar and boil rapidly until juice gives test for jelly. Pour into sterile glasses. Cover with paraffin.

To Freeze Blackberries

Freeze only sweet, plump berries with dark glossy skins. Wash berries in iced water. Discard over and under-ripe fruit. Drain berries in a colander or on absorbent paper towels.

An unsweetened pack is satisfactory. When ready to serve, partially thaw berries and add sugar or honey to taste.

A sugar pack may be made by mixing ¾ cup sugar with each quart (1 ⅓ pounds) of berries.

The Thornless Boysenberry is one of the latest creations in wine fruits. Berries are 1 ½ to 2 inches long and 1 inch thick. Flavor is a combination of Raspberry, Blackberry and Loganberry.

6.

Dewberries and Their Delightful Cousins

(Boysenberries, Youngberries and Logan Berries)

hose who live in the northern states may have a bit of an edge on gardeners in the South and Southwest when it comes to growing Raspberries, but Dewberries and their near relatives, Boysenberries and Youngberries, are of especial value to growers in mild climates. These newer varieties largely have replaced Loganberries.

Dewberries are often called "trailing Blackberries," and they are indeed kissin' cousins. But their habit of growth is very different. Whereas Blackberries for the most part grow in an erect manner, Dewberries and their relatives, Boysen and Young, trail along the ground, leisurely covering fences, nearby bushes, low-lying walls or anything else that happens to be handy. And wherever a tip bends down and becomes covered with earth, it will form a new plant. For this reason such types usually are trellised, to keep the fruit clean and make it easier to pick.

The Dew-Blackberry, a cross between Dewberries and Blackberries (Burgess) is more erect in its habit of growth, resembling its Blackberry parent in this respect. The large, delicious, Dewberry-flavored berries are borne abundantly. Thorns are much in evidence on this hybrid, but I always pick diligently—figuring I'll spend myself now and heal up later!

Lucretia Dewberry, one of the first to be developed, still is considered one of the best varieties to grow, especially in the western area. The Boysen and Youngberry types, though some-what tender, are thought to be the best for the eastern area. All of these may be trained on stakes about four feet high driven into each hill. Space these plantings in rows about six feet by six feet. Pick up canes each spring just before growth starts, wrap around the stakes and tie near the top. Old canes are best removed soon after harvesting.

Another method often used with Boysen and Youngberries is a two or three-wire trellis. This may be constructed with sturdy, well-braced end posts about 25 to 30 feet apart and a post in the row between every third or fourth plant. Treat the portion of the posts to be inserted in the ground with a wood preservative. In the spring before growth starts, but *after danger of heavy frost has passed*, divide the canes from one hill about equally and wrap around the wires running in each direction from the plant.

The number of wrappings from the top wire to the bottom wire and back up to the top will depend on the length of the canes. Sometimes the cane growth is shortened to not more than eight feet. Binder twine usually is used to tie the canes near the end.

Unlike Blackberries, Dewberry types root deeply. This enables them to withstand conditions of drought and makes them suitable for planting in the dry portions of the South and Southwest.

PLANTING IS A SIMPLE MATTER

Dewberries are planted in much the same manner as Blackberries, placing them at the same depth at which they formerly grew. They should be cut back to about six inches and spaced about six feet apart in the row. If you plan on having more than one row, space the rows eight feet apart.

Youngberries and Boysenberries have longer fruiting canes than Dewberries and for this reason they should be placed eight feet apart in the row. The big, wine-red berries of brother Boysen are especially great for jams, jellies, canning, freezing and wine. It doesn't take many plants to fill a gallon pail, for each will yield as much as two gallons during the season. To this I can testify personally, for I have finished picking many gallons so recently that my fingers still are stained with berry juice.

My garden space is limited enough that I cannot grow everything, but I have room for two Dewberry and two Boysenberry plants. Most years both ripen at almost the same time. I keep these four plants for a check on ripening dates, which may vary as much as a week or ten days from one season to the next.

The Dewberry plants were dug from wild patches, discovered several years ago by my teenage son when he was on hunting and fishing expeditions. Brought up to be berry-conscious, he dutifully reported home when he found a high-yielding area of good plants. My two plants, which are renewed from time to time, tell me when the wild berries are ripe. Then I lose no time sallying forth to pick.

My two Boysenberry plants were given to me by a friend who raises them on a commercial basis. He has many acres planted to Boysenberries, and on his land, which has good sandy loam, the yields are tremendous. Each year he throws open his plantings to people who

---

Common name: Dewberry
Botanical name: *Rubus villosus*
Soil: Deep, fertile loam.
**Nutrients: Organic matter such as compost or well-rotted
    manure.**
**Water: Moist but not over wet. Water well in summer in hot
    climates. Their deep rooting enables them to withstand
    drought in the South and Southwest.**
Spacing: 6 feet apart.
Sunlight: Full.
**When to plant: Early spring or, if obtainable, fall in
    South/Southwest.**
**When to prune: Allow to sprawl first year; train to supports
    second. Cut off old canes after fruiting and burn them,
    keeping 14 or 15 new canes for next season. Cut back to 6
    inches when transplanting.**

are permitted to "pick on the halves." For each two gallons picked, the picker receives one free, the other gallon going to the owner. Or the picker may purchase the second gallon if he so wishes. Carl and I always buy ours, and if we do not want to make up the berries right away we freeze them to use at a more convenient time. Since this acreage is about fifteen miles away, my two Boysenberry plants also serve as a check. When they are ripe I know it is time to drive to the berry ranch.

FIRST YEAR

The first season after planting, all three of these types may be allowed to sprawl on the ground, which is their natural inclination. Just let them grow. Keep down weeds by clean cultivation between the rows. Mulching is desirable but not absolutely necessary if rainfall is sufficient and the soil is well supplied with organic matter.

SECOND YEAR

When the second spring arrives it's a good idea to train the plants to some form of support, for the vines allowed to roam at will then become difficult to control. Also, if the plants are not mulched and the fruiting canes lie on the ground, the berries will get very dirty. In a wet season they even may rot before they fully ripen. To assure yourself of a good crop, get the vines up where they will have good air circulation. Corral them without delay.

As the young new canes develop, let them lie on the ground as you did the first year for the new planting. After the old canes have borne fruit, cut them off close to the ground and burn them. Then select about fourteen or fifteen canes for the next season's crop from the new canes. Limiting the number of canes will assure you of a crop of large, delicious berries.

If you live in a mild climate where winters are not severe, tie the newly-selected canes up to the wires after cutting out the old, spent

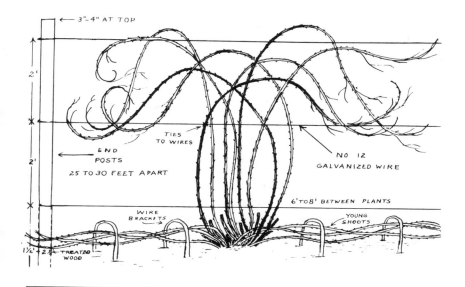

Training Dewberries, Boysenberries and Youngberries.

All three of these berries tend to grow in a rather sprawling manner and the fruiting canes will need support. They will also form unwanted tip plants if allowed to bed down. Young shoots of the present year's growth are kept on the ground and fastened with wire brackets to keep them out of the way. Fruiting canes are looped over trellis and cut off after bearing.

In the Northern states the canes kept on the ground may be covered with earth and mulch during the winter months. In the spring divide the canes from one hill about equally and wrap them around the wires running each direction from the plant.

canes. But if you live in a northern section, let the canes remain on the ground, covering them with several inches of earth. If you can get some bales of old hay, place several inches of this over the plants also.

In the spring, pull back the hay, uncover the plants, and tie them to the wires. Save the hay and use it for mulching.

FERTILIZATION

In the spring, fertilizer (in the form of compost or well-rotted manure) should be worked into the soil around each plant. Depending on where you live, March or April is a good time to do this. To keep a planting in good health and good bearing, you must always keep a feeding program going. Remember that even though they mostly stay in one place and seldom make a noise, plants are living things and they get hungry too.

FREEZING

Dewberries, Boysenberries, Youngberries and Loganberries all are of very delicate structure—another reason for keeping the fruit as clean as possible, so that much washing will not be necessary. The skins are easily broken and the juice lost. Always chill down the warm, newly-picked berries first by placing them in the refrigerator for several hours. Wash in iced water.

Select sweet, plump berries with glossy skins. After washing, drain the fruit in a colander (a few at a time—do not pack), or on absorbent towels. Over and under-ripe fruit should be discarded.

An unsweetened pack is satisfactory. Or mix ¾ cup sugar with each quart (1 ⅓ pounds) of berries.

Another way of preserving these berries is to run them through a fruit press. This is especially helpful if you have a large quantity intended for wine or jellymaking. Boysenberry wine, a clear deep violet-red, is breathtakingly beautiful and has a marvelous flavor. The jelly is the same color and may be made from the fresh berry juice or from juice which has been frozen.

I do not like to make up all my jelly at the same time—it tends to harden more with age and also lose some of its fresh-made flavor. Therefore I use empty gallon milk cartons, wash them carefully, fill with juice, staple them shut and freeze. This has the advantage also of allowing me to make jelly during the cool winter months when there is less garden work to be done.

Berry Wine
(Dewberry, Boysenberry, Youngberry, or Loganberry)

3 pounds of berries
3 pounds of sugar
20 grams of 80% lactic acid per quart of "must," the expressed
    juice of grapes or berries, before fermentation
All-purpose wine yeast
1 yeast Food tablet
7 pints of water

Directions for making wine come with wine-making kits. Or you might purchase "How To Make The Finest Wines at Home" by George Leonard Herter.

The Blueberry is a shrub of the heath family. Blueberries are often mistaken for huckleberries, which they are not, being more closely related to cranberries.

7.

Blueberries in Your Backyard

ll berries like company, but Blueberries like it the most, the reason being that Blueberries are largely self-sterile. Therefore be sure to select at least two different varieties, to avoid the risk of having your plants fail to bear. Bumblebees are the Blueberry's pollinating insect.

There are three types of Blueberries: the Highbush, the Low-bush and the Rabbiteye. The Highbush is best adapted to light sandy soils that are well blessed with organic matter. They also need copious supplies of water from a relatively high water table or the advantage of frequent rainfall.

This type usually produces best on acid type soil, and its production is somewhat restricted. It grows well in North Carolina, New Jersey, Massachusetts and Michigan, best in the eastern United States.

Just as the name would indicate, it will attain a height of six to eight feet. If its rather rigid soil requirements are met, it is exceedingly productive.

The Lowbush Blueberry, as you would expect from the name, is a much smaller growing plant, seldom taller than two feet. It is grown in the New England states where the cold winters are just too much for the Highbush type. It also does best in acid soil. Here it frequently is covered with snow through the coldest part of the winter, and this acts as a protective blanket against winter injury.

The Rabbiteye, the third type, is a native of the southeastern United States, and it is better adapted than the other two to a wide range of soil types. Where the Rabbiteye is well adapted, it is very vigorous in growth and may be even larger than the Highbush.

Dr. Frederick Coville, who began his studies in 1906, is largely credited with making it possible for homeowners to grow Blueberries. Up to the time of his experiments most home plantings were a failure, due to the fact that they were planted in soil not sufficiently acidic.

You knew I was sneaking up to something, didn't you? I am, and it's that old pH factor. Everybody who tells you how to grow Blueberries begins by stating that soil pH should be between 4.2 and 5.2. You are expected to know what they are talking about, and I think this is unfair. I didn't, so I looked it up and here is what I found. The symbol "pH" is simply a measure of soil acidity.

The pH of anything indicates its active acidity or alkalinity expressed in units. It is important, because many plants—and Blueberries in particular—will thrive only when the pH value of the soil in which they are planted is similar to their native environment, or is made to approximate it closely by the addition of organic matter.

The exact pH of your soil may be determined by a soil test. Granted that Blueberries do best in a cool, moist climate without hot, drying winds, the whole secret in growing them is soil preparation.

To make a soil acid, if it isn't already, you must add things to it and, for our purpose, organic substances are best. Manure, however, which is so valuable to promote growth for other berries, is not good for Blueberries, for it tends to make the soil alkaline. Among the best organic additives are partially decayed oak leaves and acid peat, obtainable at most gardening centers.

So if you already have soil that is naturally sandy, just mix in plenty of acid peat. If it is clay, you will need to add sand. Those who live in northern and eastern sections can grow Blueberries wherever flowers, vegetables or other fruits grow. They also will do well on meadow-type soils, which often are too wet for other crops, and the fruit yield will be especially heavy because of the added moisture.

Highbush plants in particular, when properly set out and given an annual feeding and a good mulch, may be expected to last to the lifetime of the gardener.

BEST TIME TO PLANT

Plant your Blueberries when they are dormant. This may be in spring as soon as the ground is ready, or in the fall as soon as the leaves begin to show color. Any time after that until the ground freezes too hard to be workable is all right. After planting, put on a good, heavy mulch to avoid the danger of heaving which will tear the roots and expose them to drying by sun and wind.

The best spacing for plants is six to eight feet, so the sun, which Blueberries insist upon having, can reach all parts of the bush equally. This results in better ripening of the fruit, and they will cross-fertilize satisfactorily.

Common name: Blueberry
Botanical name: *Vaccinium corymbosum* **(highbush)**
V. pensilvanicum **(***angustifolium***) (lowbush)**
Soil: Acidic (5-5.6), sandy or peaty. Draining to a depth of 18 or 20 inches.
Nutrients: Prefer humus and soft, woodsy acidic soil so much that it is almost a question of growing them organically or not growing them at all. Add liberal amounts of peaty material.
Water: Needs abundant moisture to 18 or 20 inches.
Spacing: 5 feet apart in rows 7 feet apart. Hedges: 3 feet apart.
Sunlight: Full.
When to plant: Early spring or late fall. Plant when dormant.
When to prune: Not until 4 or 5 years after set. Prune lightly (old canes and thin, bushy wood) after harvest ceases.

PLANTING AND FEEDING

It is not difficult for the home gardener to create a mini- environment. The soil should be prepared carefully, and if this is done beforehand many problems often met in Blueberry culture will not happen. A good soil mixture is composed of one pint of sulfur, one bushel of old sawdust (preferably not pine sawdust), and two bushels of sandy loam. These should be mixed thoroughly and allowed to stand for two months before using.

Dig your hole for each plant twice the size needed for the plant, and use this soil mixture for refilling the hole. The plant should be set high, so that the crown is two inches above the soil level.

The first fall after planting, the area should be heavily mulched with hay if you can get it. Put on at least five or six inches as it will pack down to about three. If you can't get hay, you can use well-decomposed sawdust, putting on no more than four inches. Sawdust tends to pack tightly, especially if fresh, and gentle rains have difficulty penetrating if it is used too liberally. What does soak in may be retained by the sawdust, greedily withholding it from the bushes.

Preferable to sawdust are other organic materials such as pine needles, garden wastes, leaves, lawn clippings, wood chips, straw and even horse manure, well mixed with bedding.

In fertilizing Blueberries it is best to be a little on the lean side, for overfeeding will produce lovely foliage but little fruit, especially if too much nitrogen is applied.

Blueberry growers frequently are advised to use aluminum sulphate, but I do not. This makes the flavor of the berries very acid and also is detrimental to the bacterial life of the soil.

The best Blueberry fertilizers contain organic sources of nitrogen, phosphorus and potash in about equal amounts. Remember, however, that established plants can take some of their nitrogen from the air and from rainfall. If they are fertilized with an all-around mixture early in the season before they leaf out they usually will receive plenty of nitrogen. Good slow-releasing sources of nitrogen include cottonseed and soybean meal, fish meal, dried blood, sheep manure, castor bean pomace, bone meal, corn gluten meal, etc.

Never add fertilizer when first planting your bushes. Wait until they have leafed out and the leaves are fully matured. Then you can stir fertilizer into the upper 3 inches of the soil. Blueberries are shallow rooted, so do not go any deeper. Water well, so that the fertilizer will become readily available. Pinch off all blossoms the first year after planting.

You may use some manure at this time if it is well decomposed, but keep it away from the crown of the plant. Actually, if you use manure, it is better to put it on when the plant is dormant, usually from November to February. To keep up the soil's mineral health, apply phosphate rock and granite dust every three or four years.

Fertilizer is best applied in small doses beginning in March and following through with another in April and the last one in May. About three handfuls, spread around in the area under the drip line and away from the main stem is about right for a three-year-old bush. As the bush grows larger it can use progressively more, until it can profit from about two pounds at nine or ten years of age.

Understanding how a Blueberry grows helps to understand its rather voracious appetite. It has to produce new feeding roots each spring. It must develop the fruit buds already set the previous year for this season's crop. And it must set buds for the coming year.

This is a very busy time indeed for the bush, for all of this must be accomplished within the short span of time from the opening of the flower buds to the final growth of the fruit. Without early fertilization and adequate rainfall during this critical time the bush simply cannot perform at its best.

BLUEBERRIES NEED LITTLE PRUNING

Blueberries are undemanding in the matter of pruning for the first few years after planting. Aside from cutting small twiggy growth off at ground level, at least by the third year, and disposing of any broken branches, there is little to be done.

What we are working toward is a bush that will produce tall, erect, strong canes directly from the root crown. Sometimes people confuse these shoots with suckers and prune them out. This is a mistake, for they are really the framework of the productive bush of coming seasons.

Plants should be pruned annually after the fourth year, doing this labor of love only when the plant is dormant in late fall, winter or early spring. At this time cut back all damaged wood to healthy, strong growth. Be sure to collect and remove all prunings. (By now, you've probably gotten very tired of listening to me repeat this but it IS necessary to keep a clean, uncluttered patch with all berry varieties. It helps to keep down disease.)

Pruning methods vary according to the kinds of Blueberries you will grow. Some types are upright, producing a number of long, slender main branches from the crown. Others are more compact and spreading in habit, forming small side branches in the heads of the bushes.

The uprights are pruned mostly by cutting out surplus main branches; the stocky, spreading types must have their small side growths thinned.

Don't ever lose sight of "why" you are pruning; it is to encourage the bushes to produce plenty of laterals, or side shoots, each year. Remember that fruit is borne only on these laterals of the previous year's growth. As with other berries, these laterals should be removed after fruiting. This will make room for the new shoots that were growing quietly while they were getting ready to bear.

As your bushes grow larger, do not try to save all the new shoots. Steel yourself. Save only those which are strong and vigorous; thin out the weaklings ruthlessly.

After your bushes reach six or seven years, older branches may not be producing good lateral growth any longer. Cut back to strong branches or cut right down to the ground. If you do this the bush will continue to renew itself and old wood will be kept off.

Young main branches will produce more vigorous laterals than old wood, so it is a good idea to remove a certain number of the main branches each year starting the third or fourth year, cutting them off close to the crown. With either upright or spreading bushes, also cut out any low-hanging branches, since the fruit which develops on these may become very dirty or even rot.

As with many other kinds of fruit, another purpose of pruning is

Blueberries will not need pruning for the first three years. The third year after planting, in early spring before growth begins, remove dead or injured branches, short and stubby branches near the ground and any old stems low in vigor.

to prevent excessive bearing. Berry bushes just don't seem to have much judgment in this regard. Left to their own devices they will bear all they can. If you don't prevent this by judicious pruning you may get lots of fruit, but it will be small and of inferior quality. Less fruit with better and bigger berries will be the result of careful pruning, and this is always more desirable.

How can you do this and do it well? Easily. Simply shorten the fruiting laterals, removing some of their fruit buds. How can you tell which are the fruit buds and which are the leaf buds? Easy again. The fruit buds are fatter than the leaf buds.

Blueberries are very apt to get carried away in the matter of setting berries and each bud will produce a cluster of nine to fourteen berries. Three or so clusters is all a lateral can bear and bear well, so harden your heart and cut back to three fruit buds. Some varieties need more restraint in this matter than others. Tip them back severely. Others grow their fruit buds on the third terminal of the laterals and with these only moderate tipping is needed.

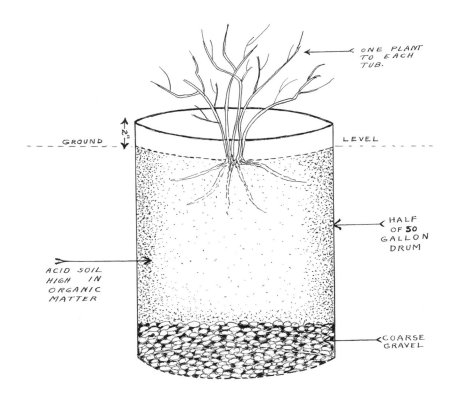

ONE PLANT
TO EACH
TUB.

GROUND   LEVEL

2"

HALF
OF **50**
GALLON
DRUM

ACID SOIL
HIGH IN
ORGANIC
MATTER

COARSE
GRAVEL

BLUEBERRIES HAVE FEW DISEASE OR INSECT TROUBLES

Blueberries, indeed, are seldom troubled by disease, and very few insects seem to bother them. The **Blueberry Maggot** and **Fruit Fly** sometimes bother, but allowing it to become a real nuisance will be mostly your own fault. Neglecting to pick fruit when it's ripe or leaving it hanging for any length of time encourages this pest. It is under these conditions that the fly will lay eggs in the over-ripe berries.

Keep berries picked and clean up all old trimmings promptly, since these provide a breeding ground for insects that may get started before you realize they are present. Dusting with rotenone may prevent insect invasion.

If you find a wilted tip on a plant, break it off promptly and burn it. This will catch the little stem borer and he will be destroyed. **Milky Spore** disease is being used as a defense against **Japanese Beetles**, which sometimes prove a nuisance with Blueberries in some parts of the country. And if you live in an area where **rabbits** are troublesome during the winter months, a low fence should be placed around your plants.

All in all I think the **birds** are more troublesome than anything else. They will eat all your berries if you do not take steps to provide adequate protection—or else plant enough for both of you. Cheese-cloth once was used to cover the fruiting bushes, but today nylon nets are much more practical. And the mesh in dark colors looks much better on the bushes.

Good health in Blueberries often is indicated by the color of the leaves. They should be large and dark green, though the intensity of the color may vary with different varieties. A yellow-green color may indicate nitrogen deficiency or a wet, poorly drained soil. Healthy, strong-growing bushes always are more resistant to disease and insect troubles. The yellow-green color also may indicate an alkaline soil, which may be associated with iron deficiency. If you suspect this, try adding iron chelates, placing them in a ring around the drip line of the bush and working them lightly into the soil. They are available at garden centers. Used for the same purposes is a new product, GU-49, which contains a minimum of 49% micron-sized total iron and other nutritional trace elements. It breaks down readily so that it is easily assimilated by the plant roots.

BLUEBERRIES AS TUB PLANTS

It's not difficult to get hooked on Blueberries, for few berries are more tasty for eating fresh or baking into pies and muffins, and they are easily frozen. But their exacting soil, climate and cultural requirements cause many home gardeners to feel they cannot grow them success-fully—or they have tried and failed.

Here is a way that may enable you to enjoy at least a few plants, and it's especially helpful if your soil is not acid enough for Blueber-ries—about 5 pH. You simply create a mini-garden by growing them in tubs buried in your garden. Halves of 50-gallon drums, with drainage holes cut in the bottom, work well.

It usually will be necessary first to burn out any residues that might injure the plants. You can rent blow torches that will sear the tub. This done, bury the drum in a sunny location, leaving 1 to 2 inches of the rim above ground level. Fill the bottom part of the tub with several inches of coarse gravel to afford drainage and then put acid soil (pH 4.2 to 5.2) high in organic matter (use acid peat moss to mix with sand if you can't get anything else) in the drum. Set one Blueberry plant in each tub.

MAKE A BLUEBERRY HEDGE

Blueberries make beautiful hedge plants. The highbush type is especially useful for screening an undesirable view. In the spring you will have white- or pink-tinted flowers, somewhat resembling the Lily-of-the-Valley in shape. Dense green foliage covers the bushes, turning to scarlet and crimson in the fall, and even in winter the red twigs are lovely against the snow. And, of course, the main dish in this gourmet's color delight are the blue Blueberries.

To have your hedge do its job satisfactorily, the plants should be placed closer together, three feet apart in the row is a good distance. Dig your holes two to three feet wide and about eighteen inches deep. Let them be large enough so you can spread the roots out flat rather than letting them hang down. Plant the blueberry bushes, fill the holes with good humus soil, tamp and water. Mulch to keep down weeds and retain moisture.

If your plants arrive in a dried out condition be sure to soak them well in a bucket of water before planting. Carry the bucket with you as you go and never let the roots have a chance to dry out.

Fertilize the bushes each spring with a *light* application of cotton-seed meal. Prune about the third year by cutting out weak canes and heading back any others that may have gotten too long, so the hedge will retain its desirable shape. In summer, water thoroughly if there is a prolonged dry spell. To ensure fruiting and to have plants about the same height, good choices are Blueray, Coville and Jersey.

In late summer pick the berries and enjoy them!

GROWING SOUTHERN BLUEBERRIES

The Rabbiteye Blueberry, being a southern native, does not need the long, cold dormant season required by the northern types for their best development. The Rabbiteye needs a minimum of only about 500 hours below 45°F, where the northern highbush types must have from 700 to 1,000 hours below 45 degrees. Because of their drought and heat-resisting character, Rabbiteyes will grow in many areas, including the Texas coastal plain, and they also will tolerate neutral soils that are prepared and fertilized much as those needed by camellias and azaleas. Moist, sandy loams with humus are ideal.

There is another bonus in growing the Rabbiteye: they are very vigorous plants little troubled by pests and disease. They also are attractive at all seasons and stages of growth. Pretty for hedges, you can plant them three or four feet apart and they will make a three foot wide border when they mature.

In early spring the bushes are covered with delightful white blossoms. They set a bountiful crop of delicious, good-flavored berries over a midsummer season of six to eight weeks.

Rabbiteye plants have a fibrous, rather shallow root system, so do not cultivate deeply. The most satisfactory methods of weed control are hand hoeing and mulching. Put on a layer of wood chips, sawdust or some similar material to keep down weeds and conserve moisture.

To get your soil down to the acid pH desirable, it will be necessary to incorporate a great deal of old sawdust, aged pine needles, oak leaves,

Blueberry Propagation

Blueberries usually are grown from hardwood cuttings, but it is a job best left to the experts, for the cuttings are more difficult to root than most other plants. The home gardener will find it more satisfactory to buy plants from nurseries specializing in Blueberry propagation.

etc. in your soil. Sphagnum peat moss also is good to add. Here is where the use of a rotary tiller will make your job easier and help you to avoid the back-breaking work of mixing. It can do the job easily.

The Rabbiteyes require very little pruning, and about all that is necessary is the removal of dead or damaged wood. Generally speaking, heading back of the shoots should be avoided, for this will cause excessive loss of the fruit buds, which form near the ends of the branches.

If you must plant your Blueberries on a very high, dry soil where they are subject to periods of prolonged drought, you will need to water them during such times. Though they do not need as much moisture as the highbush varieties, they cannot stand being dry too long.

Always provide light shade for your young transplants. Pine boughs are good if you have them. But once established, Rabbit-eyes like other Blueberry types will grow best in full sun, though light shade is tolerated.

Use compost for feeding, well-decomposed manure or other organic fertilizers. The plants are very sensitive to strong chemical fertilizers and may be killed by excessive applications. Homebell is a vigorous, upright-growing plant which produces abundantly. The fruit is medium to large in size, good blue in color and has excellent quality. It is best for home use, as it does not ship as well as some of the others. It is also ideal as a pollinater. Early to mid-season.

Tifblue is one of the finest of southern Blueberries, making an upright bush with large, light blue berries of extremely high quality and sweetness. One of the most widely planted, it ripens early to mid-season.

Menditoo, a variety from the North Carolina Experiment Station, has large, round, dark blue fruit also of good quality. It ripens from late June through July. Plants have medium vigor and height and are somewhat spreading.

Rabbiteye Blueberries have proven satisfactory to grow in all the Gulf States and also as far north as Maryland and the Missouri Ozarks.

Scientists are now also working to develop the southern native highbush Blueberry. These have similar requirements to the Rabbiteye but need a light, well-drained, moist soil for best growth.

BLUEBERRY VARIETIES

Until about sixty or so years ago Blueberries were considered mostly a "wild" crop. But we are lucky in that improved varieties, some representing complicated crosses, now are available. They are larger and tastier than ever.

Here is what you should look for in Blueberries. You will want flavor, firmness and productivity. They should have a light blue color and a good scar (the end that severs from the bush), while the blossom end should be smooth. A tall, upright-spreading, V-shape is desirable, as well as increased resistance to pests and diseases. You will want both early and late fruiting varieties and, of course, large berries as long as quality also stays high.

In the Highbush variety Earliblue is one of the first to ripen followed shortly thereafter by Bluecrop and Blueray. Mid-season varieties include Herbert and Collins. Penberton, planted with Herbert or Blueray is an excellent producer. Coville is usually the last to ripen and is a really big berry, extending the season well beyond other types. Darrow, a promising newcomer, is also late. Jersey, Rancocas and Rubel are also excellent varieties for the North.

The older varieties Cabot and Greenfield are still good Lowbush types.

In eastern North Carolina and further south through the Gulf Coastal States and westward to Arkansas, the Rabbiteye varieties are best—such berries as Myers, Hagood and Black Giant. There is even a Blueberry that will do well in Florida: Bluegem, a light blue beauty, will give excellent yields if planted with Woodard.

RECIPES

To Freeze Blueberries

When that happy day arrives that you have more Blueberries than you want to eat or make up into pies, think about freezing them. Clean, fresh-picked berries need not be washed. Place them on shallow trays and freeze. This way when you sack them up you can take out as many or as few as you like at a time. They will roll out like little marbles and not stick together in an unmanageable mass.

Blueberry Muffins

2 cups flour
3 teaspoons baking powder
½ teaspoon salt
3 tablespoons sugar
3 tablespoons cooking oil
½ cup milk
½ cup Blueberries
1 egg, well beaten

Sift flour, measure, and sift with salt and baking powder. Combine egg, shortening, sugar, Blueberries and milk. Add dry ingredients. Mix only until blended. Fill well-oiled muffin tins ⅔ full. Bake in hot oven (425°F) 12 to 15 minutes.

Blueberry Jam

4 ½ cups Blueberries
Juice of one lemon
1 bottle fruit pectin
7 cups sugar
Grated rind of ½ lemon

Wash Blueberries thoroughly. Crush. Add lemon juice. Add grated rind of ½ lemon. Add sugar. Mix thoroughly. Heat rapidly to full rolling boil. Stir constantly both before and while boiling. Boil hard 2 minutes. Remove from fire and stir in fruit pectin. Skim if necessary. Pour in sterile glasses and seal.

Blueberry Triangles

1 shortcake recipe (see page 00)
2 tablespoons cooking oil
1 cup Blueberries
¼ cup sugar
1 tablespoon flour
1 tablespoon lemon juice

Prepare shortcake dough. Roll out ¼ inch thick and brush with cooking oil. Cut into 3-inch squares.

Mix sugar, flour, Blueberries and lemon juice together. Place a large tablespoon of mixture on each square. Moisten edges of dough with water and fold over to form a triangle. Seal edges by pressing with tines of a fork. Prick. Place on oiled 10 x 14 cookie sheet and bake in oven preheated to 400°F. Baking time 20 to 25 minutes. Serve hot with cream.

Blueberry-Apple Butter

8 cups apple pulp
8 cups Blueberries
8 cups sugar
1 tablespoon allspice
1 tablespoon lemon juice

Slice apples, add small amount of water and cook covered until soft. Press through food mill. Measure. Mix apple pulp, Blueberries, sugar, spice and lemon juice. Boil until mixture thickens. Pour, hot, into sterilized jars. Process 10 minutes in boiling-water bath.

Gooseberries (top)
Currants (bottom)

8.

Gooseberries and Currants Regain Their Self-Respect

nce upon a time these two bush fruits were very popular, and deservedly so, for there are few things more delectable than Currant jelly or Gooseberry pie or sauce.

Then came the awful discovery that these two fine fruits, innocent in themselves, serve as hosts to one stage of the White Pine Blister Rust, a very serious disease of white pine trees.

Currants and Gooseberries now may be shipped into almost every state: Gooseberries cannot be shipped to Delaware, Maine, New Hampshire, New Jersey, Vermont, and West Virginia; Currants cannot be shipped to Delaware, Maine, North Carolina, New Hampshire, New Jersey, Vermont, and West Virginia. Until recently growers could ship them only into twenty-four states, so now it is possible for more people to enjoy these tasty berries in their home gardens.

Gooseberries and Currants both are very worthwhile if you can grow them, and because their culture is similar, we'll consider them together.

They are heavy annual yielders and winter hardy anywhere in the northern United States, and they also may be grown in Canada which has no law against them. Though easy to grow in the cooler parts of the United States, they

Common name: Gooseberry
Botanical name: *Ribes grossularia* and hybrids.
Soil: Any rich, well-drained garden soil.
Nutrients: Heavy feeders; require supplemental potash.
Water: Keep soil moist throughout growing season.
Spacing: 4 to 6 feet between plants.
When to plant: Fall or very early spring.
When to prune: Midwinter.

are not happy in the South, though Gooseberries grow a bit farther south than Currants.

Even in the North they will do best in cool, moist, partially-shaded locations. The north or east side of a building, fence or arbor may provide these conditions best.

Plants are easiest to secure in the spring, but may be set in fall as well. Spring planting should be done early, before the buds begin to grow. Purchase vigorous, well-rooted, one-year-old plants. At planting time prune off any damaged roots and cut the tops back to ten inches. Set the plants with the lower branches just a little below the soil level, which will encourage them to develop in bush form. Space the plants four to six feet apart and, if you have more than one row, make these six to eight feet apart.

HUNGRY PLANTS

Currants and Gooseberries are heavy feeders. If you placed generous amounts of manure (well-composted) in the soil when you planted them, you won't need to give them any more, however, the first year. After that you should work some extra compost into the soil around the plants each fall. Never dig this in too deeply for they have shallow root systems.

The first year after planting you must make sure that the soil does not dry out. Water when necessary, and keep down weeds with surface cultivation, or mulch with straw, corncobs, lawn clippings or sawdust.

RESTRAIN THEIR ENTHUSIASM

As if they felt unjustly accused and wanted to clear their names by proving how strong and healthy they are, Currants and Gooseberries have exceedingly vigorous growth. And, indeed, the disease, which they harbor through an intermediary growth stage, does not affect them at all. It is fatal only to the White Pines.

Annual pruning is required in both Currants and Gooseberries for maximum production. The red and white Currants, as well as the Gooseberries, develop fruit from buds at the base of one-year wood and from spurs on older wood.

The older wood will become progressively less fruitful, and canes

Common name: Currant
Botanical name: *Ribes* **spp.**
Soil: Well-drained with ample organic matter.
Nutrients: Well-rotted manure or compost.
Water: Keep moist but not over-wet. Do not let them dry out.
Spacing: Space canes 6 to 8 inches apart, always removing the
oldest, saving the larger stocky ones for fruiting. Space
permanent plants 4 feet apart in rows 6 feet apart.
Sunlight: Full in a well-ventilated area; can take partial shade.
When to plant: Early spring, but fall is also okay.
When to prune: A little should be done each year, based on
fruiting habit. The best fruit is borne on canes 2 to 4 years
old. Cut old canes to stimulate new ones to bear for 3 or
more years.

which have borne for three years usually become unproductive. Judicious pruning consists mostly of selecting the proper type of fruiting wood.

All pruning should be done in early spring when the plants are dormant. After the first year, remove the weak shoots, leaving six to eight strong branches.

The second year, remove all but four or five of the two-year-old branches and four or five of the one-year-old branches.

On the third and subsequent years, leave four or five three-year old branches, four or five two year-old branches, and four or five one-year-old branches (of the previous season's growth).

This program will add up to twelve to fifteen branches for each plant.

When pruning, remove branches that hang low or lie on the ground. Also try to remove any weak branches in the center of the bush, to prevent the plant from becoming too dense.

CURRANT VARIETIES

Red Lake is still one of the most popular varieties. It is a vigorous and productive bush, yielding tremendous quantities of high quality large red Currants. It is very hardy and often will bear the first year after planting.

Improved Perfection (Gurney's) has even larger berries borne so profusely they almost cover the plant.

Wilder also is vigorous and productive, with large, red, firm, juicy fruit. It has the added advantage that the fruit hang longer after ripening.

GOOSEBERRY VARIETIES

Pixwell, one of the best, is a very dark red upon ripening. The berries hang well away from the thorns, making picking much more comfortable. The crops produced are large and the berries big. Pixwell will grow and thrive in any climate where Gooseberries can be grown.

Welcome, a University of Minnesota introduction, is perfectly delicious for pies, jelly or jam. The attractive, light green fruit turns pink when fully ripe.

In America most people seem to prefer to harvest Gooseberries in a green state, but the English prefer to let them ripen to a dark plum color and serve them with sugar and cream, as we do so many of our other berries. For pies Gooseberries are good either way, green or well ripened. You will need to use a little more sugar for the green berries and less for those fully mature.

PESTS ARE FEW

Currants and Gooseberries require very little in the way of a spray program. The most troublesome insects for both fruits are the **Currant aphid**, which causes bright red, cupped, distorted or wrinkled areas on the leaves; the imported **Currant worm**, about one inch long and greenish in color with black spots, and which feeds on the edges of the leaves; and the **cane borer**. This last is a worm about one-half inch long which burrows the entire length of the cane and causes dwarfing of the plant. **Scale** insects, whose small, grayish bodies can be seen on the bark, occasionally are troublesome.

To control scale, mites and aphids, use a dormant spray. This consists of using plant spray oil, 6 tablespoons to the gallon, *just before* the buds start to swell. *Do no use this later.* Sometimes a special spray— wettable sulfur at two tablespoons per gallon—may be needed to control powdery mildew.

When pruning Currants and Gooseberries during the dormant season look for the signs of cane borers: hollow canes with black centers. Cut all such canes out entirely and burn them without delay.

Keeping the plants from becoming too dense will reduce damage from diseases, as this will permit good air circulation and quick drying of the foliage. If any Currant canes suddenly wilt and die during the growing season it may be caused by **cane blight**. Cut these canes out immediately and burn them or remove them from the premises.

RECIPES

Green Gooseberry pie is a treat few people have ever tasted. And here again the home garden grower-of-Gooseberries has a decided advantage, for they should be picked when half grown. Furthermore this early picking of a portion of the crop thins out the berries, and the fully ripened fruits will be much larger. I've got kitchen-plans for those, too, but here is the one for Gooseberry pie.

Green Goosebery Pie

Have ready a 9-inch pie plate. Preheat oven to 425°F. Baking time 45–55 minutes.

4 cups green Gooseberries (remove stems and blossom ends)
2 cups sugar
5 tablespoons all-purpose flour
¼ teaspoon salt
3 tablespoons butter or margarine

Make pastry for double-crust pie. Place Gooseberries in mixing bowl. Blend together sugar, flour and salt and mix through the Gooseberries. Pour into pastry-lined pie plate. Dot with butter or margarine. Add top crust.

Spiced Gooseberries

5 pounds *ripe* Gooseberries
4 pounds brown sugar
2 cups vinegar
2 tablespoons cloves
3 teaspoons cinnamon
2 teaspoons allspice

Wash and pick over Gooseberries. Combine with remaining ingredients. Cook slowly until mixture thickens slightly. Pour into sterile glasses and seal.

Gooseberry Relish

Gooseberry lovers everywhere will enjoy this unusual relish!

5 cups Gooseberries
1 cup brown sugar
1 ½ cups raisins
1 onion, peeled and chopped
3 tablespoons salt (level)
¼ teaspoon Cayenne pepper
1 teaspoon dry mustard
1 teaspoon ginger
1 teaspoon turmeric
1 quart vinegar

Chop Gooseberries and raisins and mix with onion. Add other ingredients. Heat slowly to boiling; simmer ¾ hour. Stir often. Rub through coarse sieve. Reheat to bubbling and pour into pint fruit jars.

Currant Conserve

5 pounds Currants
2 pounds seeded raisins
5 pounds sugar
3 oranges
2 cups raspberry juice

Wash and stem Currants. Add raisins, raspberry juice, sugar and oranges which have been seeded and chopped. Simmer slowly, stirring constantly, until juice sheets from spoon. You will have a conserve that is almost sinfully delicious. "Sheeting from spoon" test for the jellying point: As the boiling mass nears this, it will drop from the side of the spoon in two drops which run together and slide off in a flake or sheet from the side of the spoon. When this point is reached the mixture should be removed from the heat at once.

Currant Jelly

Pick over Currants. Do not remove stems. Wash and drain. Place in preserving kettle. Mash with potato masher. Add ½ cup water to about 2 quarts of fruit. Bring to a boil and simmer until Currants appear white. Strain through a jelly bag. Measure juice. To two cups juice add ¾ cup honey and ¾ cup sugar. Cook only about 4 cupfuls of juice at a time. Stir until sugar dissolves. Cook until two drops run together and "sheet" off spoon. Pour into hot, sterile glasses and seal.

Currant Treasures

¾ cup shortening
2 ½ cups cake flour
½ teaspoon salt
½ teaspoon cinnamon
¾ cup sugar
1 teaspoon baking powder
1 teaspoon grated lemon rind
1 cup chopped nuts
1 cup currants
2 eggs, well beaten
6 tablespoons milk

Cream shortening. Sift flour, salt, cinnamon, sugar and baking powder together. Cut into shortening with dough blender until consistency of coarse cornmeal. Add remaining ingredients and blend thoroughly. Drop from a teaspoon onto a nonstick cookie sheet. Bake in a moderate oven, 350°F, about 15 minutes. Makes 7 dozen cookies.

9.

Grapes Are
Natural Swingers

ack in 1909, T. V. Munson, a man who knew a lot about Grapes, wrote this advice: "Grapes in trees are little bothered by rot and mildew. It is when the vines are held down on trellis near the ground in dense mass that these diseases attack the worst."

That, in a nutshell (or a grape skin), has been my observation, too. And I came to the conclusion a long time ago that one of the most important, if not *the* most important, things in growing good Grapes is good air circulation.

Not every gardener has the perfect spot to grow everything under ideal conditions, and sometimes must make do with what he has. The only place I have for growing Grapes is around my garden fence, which runs north and south. Fortunately there is a high terrace just outside the fence and the air circulation is good.

Even so the white Grapes which tended to bear heavily, one year became infected with brown rot and all the Grapes along the fences were a loss. But one eager branch escaped upward and grew into a big hackberry tree just outside the fence. It was totally unaffected by the humid conditions which ruined the others, the big clusters of white Grapes high up in the tree were beautiful to behold, even though I had to harvest them with a ladder.

The sunlight on the leaves and not on the Grapes determines whether or not Grapes can be grown and ripened. To obtain perfect bunches of Grapes, growers sometimes place them in bags. When positioning your Grape vines, always try to avoid shade from buildings and trees.

Grapes, like strawberries, can be grown in almost every section of our country if we are careful to choose adaptable varieties. They come in a wide range of flavors, can be used for juice and jelly, and the trellises and arbors on which they are grown can be most attractive parts of our gardens or home grounds, creating pleasant shaded areas that are useful in landscape planning and for screening off undesirable views.

When to Plant

Grapes should be planted just as soon as the soil can be prepared in early spring. Cut off any broken or too-long roots, so that they can be spread evenly in the planting hole. And be sure to set the plant slightly deeper than it grew in the nursery, arranging the roots so they are not bunched together.

Prune the plants to a single stem with two buds. A shoot will grow from each of the buds left on the young plant. If you have not yet constructed a trellis, tie the most vigorous shoot when it becomes long enough to need training to a stake four to five feet high.

Common name: Grape
Botanical name: *Vitis*
Soil: Coarse and deeply dug for excellent drainage, with a fair
amount—not too much—organic matter. Acidic (5 to 5.75).
Nutrients: Grapes do well with a rich mulch but should not be
over-fertilized.
Water: In hot summers keep plants moist but not over-wet.
Spacing: 6 to 8 feet apart in rows 9 feet apart.
Sunlight: Full.
When to plant: Early to midspring.
When to prune: Leave 4 canes on each plant per year to
produce fruit.

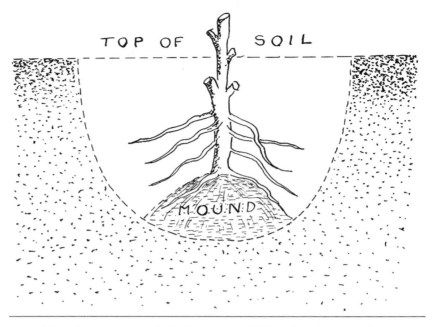

Plant Grape vine in bed of well-prepared soil. Open hole deep enough to plant vine as deep as it set in the nursery. Make a small mound in center of bottom of hole. Set plant, letting roots slant downward. Fill in with soil. Press firmly. Leave a loose layer of earth.

If you are planting in a row, space your plants eight feet apart and if a second row is planned this should also be eight feet from the first. At planting time you should take the young Grape vines to the site with their roots in a bucket of water or wrapped in moist buriap.

Dig your hole a foot or so deeper than you wish the plant to be set, and then throw in some topsoil. Place the roots in the hole and add more soil. Move the plant gently up and down to get the dirt around its roots. Continue filling the hole, tamping with your foot. Water well when you have completed the planting, so that no air pockets will be left.

Cultivate your young plants the first year. Then, after the vines become established, they can be mulched with straw, leaves or ground corncobs.

If you can get well-decomposed manure, it is the best fertilizer that can be used for Grapes, applied at the rate of one bushel around each mature vine. You also can use dry rabbit or poultry manure, but use less—five pounds will be sufficient, for these types sometimes are called the "hot" manures.

Spread the manure during late winter or early spring in a four-foot circle around each plant. Never bring it closer than one foot from the vine. Fertilizer should not be applied if the vines already are making excessive growth, for such growth will not mature properly and may be injured during the winter, especially in northern sections. Moderate growth of the canes, which mature early, is preferred.

If allowed to go unpruned, grapevines may spread as much as a hundred feet. The main stem of the vine is called the **trunk** and the main branches the **arms**. The soft green growths of the current season are called the **shoots**. When the shoots become brown and lose their leaves in the fall they are called **canes**. Grapevines develop leaves that are thin, rounded, divided and usually quite large, while the fruit are classified botanically as **berries**. They may contain one to four seeds, or be entirely seedless.

Since growing Grapes in trees is impractical for most of us (even though it may be desirable from the Grape's point of view), we must construct something to support the vines. This may be a trellis or arbor, and whichever is decided on, build it before the spring following the first growing season.

Grapes live a long time, under ideal conditions sometimes for hundreds of years. Since a planting is more or less permanent you should make the support trellis sturdy enough to last at least 20 years, remembering it must be strong enough to bear the considerable weight of mature vines and a full crop of Grapes.

The home gardener who probably will have room only for one row of grapes can still use the four-cane Kniffin training system, which is most widely used today. To build this, two wires approximately three feet apart are supported by posts that are set about twenty feet apart in the row. Galvanized wire of No. 9 or No. 10 gauge is suggested.

The lower wire should be two and one-half feet above the ground, and the top wire about five and one-half feet.

Metal or durable wood posts (of cedar, locust or white oak) should be three inches in diameter at the top and eight to eight and one-half feet long. They should be set two and one-half to three feet in the ground, or four feet in heavy frost country. Heavier posts should be used for the ends of the trellis. The end posts should be five to six inches or more in diameter at the top, and they should be nine feet long so that they can be set a full three feet deep (or four in cold climates). The end posts should be well braced to keep the trellis wires from sagging.

If you don't have room for a separate row of Grapes, try my system of growing them on your garden fence. Or build yourself a Grape arbor as a part of your landscape plan. The shading provided by vines growing over the arbor may be as valuable to you as the fruit crop.

PRUNING YOUNG VINES

I have already told you how to prune your Grapes at planting time, but Grape vines are so full of enthusiasm and eagerness to grow that this is only the beginning—never let them get the upper hand; let them know from the very start that you are the boss. An annual pruning is essential and will maintain them in good health and vigor.

The fruit clusters form from buds on one-year-old canes. The vines must be pruned in order to encourage vigorous canes to develop, to eliminate unproductive old canes, to train fruiting canes, and to limit the number of buds on the vine.

Training the vine to use the four-cane Kniffin system is easy. Simply train the mature vine to a two-wire vertical trellis. Select a permanent trunk and four one-year-old fruiting canes which are supported by the trellis.

Prune after the coldest part of winter is past and before the buds start to swell. In Oklahoma we often do this in January if the weather is mild, but farther north February or early March may be more desirable (but do not prune during the summer months). Grapes, incidentally, do not require direct sunlight to ripen and develop their full color.

You already have pruned at planting time, so the second year in early spring, while the vine is still dormant, prune off all but the strongest cane. Tie the cane tightly to the top wire of the trellis or to

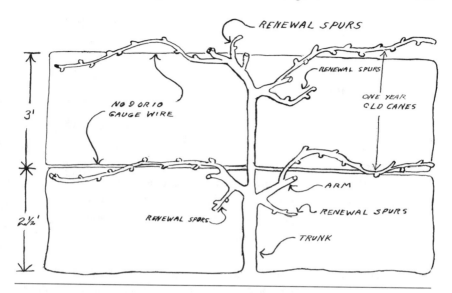

Three-year-old Grapevine trained in Four-Arm Kniffen system, a system that gives good production and requires little summer tying.

the lower wire if it is not long enough to reach the top wire. This is the main cane and will form the permanent trunk.

During the second growing season also remove any shoots that develop below the lower wire. I know this will be hard to do, but you also must remove flower clusters. During the second year, the main trunk should reach the top trellis wire, and some short lateral canes probably will develop.

Third year: if one to four strong lateral canes developed during the second year, train them to the trellis wires. If not, cut the vine back to a single vertical trunk. But in either case leave two buds (in Grape language called **renewal spurs**), on each of two shoots near the lower and upper trellis wires. Fruiting canes for the next season grow from these buds.

During the third summer you will find numerous lateral canes developing and these should bear a good crop during the fourth year. If any laterals developed during the second year, you may get a few

Grapes in the third year, or you may get some from buds on the upper part of the main trunk.

MATURE VINES

After the third year most Grapes can be treated as mature vines. But don't let up on your pruning, or the Grapes will grow in a tangle all over the place.

In early spring you should prune the vine to four lateral canes, each with six to twelve buds arising from the main trunk. Each of these buds is capable of producing two or three clusters of Grapes. Leave two renewal spurs near the main trunk for future fruiting canes at each trellis wire. And then remove all other growths.

Select canes of medium size for the lateral fruiting canes. They should be ¼ to ⅓ of an inch in diameter, straight and preferably unbranched. Do not select any canes smaller than this and don't be tempted to select long, heavy, vigorous canes. These are called "bull canes" and for a number of reasons are not the best choices.

Having made your choice, train one cane each way on the trellis wires. These lateral canes should originate from the main trunk or as near to it as possible on the arms.

After pruning, loop or spiral the canes over the supporting wires and tie with binder twine.

A Grape vine in full vigor can support forty-five to sixty buds. Twelve to fifteen buds at the most on each lateral cane may be left on vines which grew well the previous year. Actually, to prune properly, eighty to ninety percent of the wood must be removed. Most gardeners do not prune severely enough.

The Kniffin system works particulariy well with American Bunch Grapes, the type most widely planted in home gardens. American Bunch Grapes come in three different colors: red, white (actually a creamy white or yellow) and blue (sometimes so dark as to be almost black).

The red Grapes include Catawba, Delaware and a new Grape (which I am now growing here in Oklahoma with most happy results),

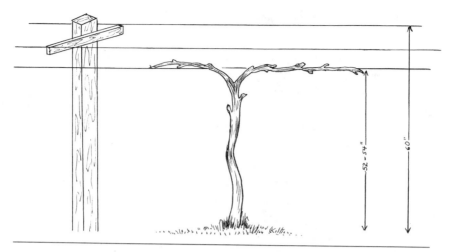

The Munson system is often used in home plantings. It is particularly suitable for humid climates, because the fruit is produced high above the ground where it is less subject to injury by diseases. Train the vine to a single trunk extending to the lower wire. After the second growing season (during the dormant period), prune to two or more canes (arms) and two renewal spurs. Tie the arms along the lower wire. As the shoots develop the next growing season, distribute then over the upper wires, allowing then to hang down. Each winter, replace the arms with canes from the renewal spurs, and leave new renewal spurs.

Burgess Red Giant. This Grape has perfectly round fruits which measure up to one and one-eighth inches in diameter and three and one-half inches in circumference. They are a deep, dark, plum-red, and just about everything that Grapes do well, Red Giant does better!

In the white class Niagara is still one of the leading varieties, with Golden Muscat a close runner-up. Interlaken Seedless is a California type that is hardy in the North and it is doing well for me even here in hot, dry, southern Oklahoma.

Blues still include the old time favorite Concord, of which there is now a seedless type, and Fredonia. Beta, an early variety, is as hardy as a wild Grape, very dependable and sets big crops every year. My favorite of these three is Fredonia, which I prize greatly for jelly and wine. Also it ripens about three weeks before Concord, giving me a longer picking season.

TRAINING AND PRUNING ARBORS

The home owner may not have room for Grapes in his or her garden or even have a fence to grow them on as I have. Planning to use them as part of the overall landscape picture is the answer. Grapes are so beautiful, so highly decorative, so exceedingly graceful in or out of fruit that they look well anywhere.

Supports in an endless variety of sizes and designs may be used. Grapes may be trained up the side of a wall or to grow on a low roof. However you plan to grow them, remember Grapes live a long time, so use support materials that will need a minimum of repair.

Allow at least eight feet between vines. Spacing too closely will not allow good air circulation and pruning will be difficult.

When you are planning your planting also keep in mind that one of your reasons for growing Grapes is to combine production of the delicious fruit with affording a shade or screen.

THE FAN SYSTEM

The Kniffin system also may be used for training Grape vines on fences or walls and it still is one of the easiest methods for purposes of maintenance. If the number of canes is increased from four to six by adding a cane on each side between the upper and lower canes, the result will be a more uniform cover.

Another method called the "fan system" is very good, too, on fences and walls. Instead of training the vine to a single trunk, it is allowed to branch a short distance above the ground. The branches then are trained into a fan-shaped arrangement.

Branching is encouraged in the fan-shape by pinching off the tip of the young vine during the growing season at the point where branching is wanted. The side branches that develop also may be pinched to obtain more stems.

Several more stems needed to develop the entire framework are unlikely the first season of training. But be patient. Once you have developed the "skeleton" of the fan, it should be maintained annually by selecting the needed one-year-old canes and leaving renewal spurs

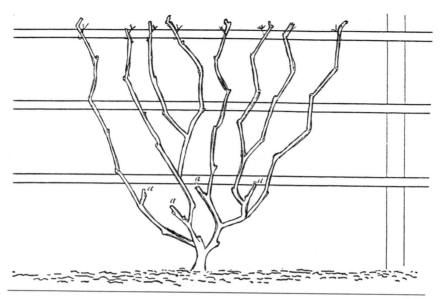

The Fan System—good for fences and walls. Induce branching by pinching off the tip of the young vine during the growing season at points where branching is desirable. You may also pinch any side branches that develop to obtain more stems and positions. Do not try to develop the entire framework the first season. Once you achieve the final form, maintain by annually selecting desired one-year-old canes and leaving renewal spurs (a) at the base of each branch to produce fruiting canes for the next season.

at the base of each branch. These will produce fruiting canes for the following year as with other systems.

The number of fruiting canes you will leave and the number of buds must be determined by the vigor of your individual vine. If the cane growth is rank and the fruit is poor, leave more buds. Cane growth that is sparse, with numerous bunches of Grapes that ripen unevenly, should be cut back more severely.

When overhead cover on arbors or summer houses is the object, the permanent single trunk is carried up along the top of the structure. Each year one-year-old canes three to four feet long are allowed to grow at intervals of two to three feet along this permanent trunk.

Renewal spurs of two or three buds are distributed along the trunk in the same way. These will produce new fruiting wood from

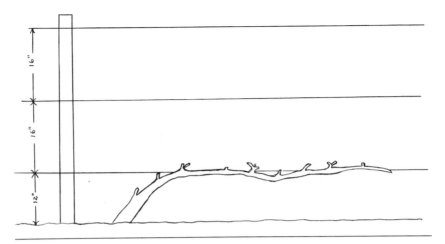

Use the Modified Chautaugua system where tender varieties of Grapes require winter protection: After the first growing season (during dormant period), select a large, vigorous cane for the trunk. Cut back to about 30 inches and remove all other canes. Lay the trunk on the ground and cover it with 6 to 8 inches of earth to protect it during the winter.

In the spring, uncover the trunk and tie the end of it at an angle to bottom wire of trellis. As shoots develop from the trunk during the summer, tie them to upper wires.

After the second growing season (during dormant period), untie the vine from the trellis. Prune the current season's growth to short spurs, two buds long, but retain the cane nearest the tip of the trunk to form an extension to the trunk. Lay the vine on the ground and cover it with soil. Repeat procedure each growing season thereafter. Vine trunk may be extended to about 7 feet.

which a selection of one-year-old canes may be made the following year.

This type of pruning will develop the shade which is part of your objective, but it is not severe enough to produce the best fruit production.

Just as with other systems, the amount of bearing wood must be adjusted to the vigor of the vine. But remember, failure to prune heavily enough will soon result in the arbor becoming a tangled mass of vines, something many home gardeners allow to happen all too frequently. Busy with other matters they forget to prune at the proper time or do not prune heavily enough. If you build an arbor don't let this happen to you.

EUROPEAN OR CALIFORNIA GRAPES

These are the most important Grapes for commercial production, but their growth is limited mainly to California and some sections of Arizona. These varieties account for more than 90 percent of the Grapes produced commercially, but the area where they can be grown successfully is very restricted. The Lakemont Seedless (Stark Bros.) is an offspring of the famous California Thompson Seedless and is well adapted to home gardens in the South. Interlaken, previously mentioned, is a California type which may be grown successfully in the North.

MUSCADINE GRAPES

Muscadines (*Vitis rotundifolia*) are native to the southeastern United States. They do well under the high temperatures and humidity found in that area. The Muscadine is sometimes called the Scuppernong, and many know it by that name. It is more resistant than most to drought conditions and also to disease. Under favorable conditions the vines are very long-lived, but they are not hardy in the northern United States because of the low temperature conditions that prevail there.

Varities include Higgins (Burgess), a very large white Grape which requires a pollenizer and Hunt, also requiring a pollenizer, but the very best all-purpose black variety, early and very productive. Noble (Burgess), a dark pollenizer, is very productive, bearing large clusters of medium-size fruit that do not shatter easily and that ripens midseason. Noble will pollenize both dark and white vines without affecting the flavor or color. Carlos is an attractive bronze high-yielder and is self-pollinating. Excellent for eating fresh. The fruit hangs well on the vines. Albemarle, Magnolia, Roanoke, Southland, Magoon and Cowart are self-pollinating. Female plants needing pollination are Nevermiss Scuppernong, Creek and Higgins Bronze.

Where Can You Grow Muscadine Grapes?

In many parts of the country the home gardener will find this Grape one of the easiest fruits to produce, especially in the Southeast because it has a high level of resistance to disease and—*good, good, good*—it does *not* require spraying.

The important thing to remember with Muscadines, however, is that several varieties must be planted together to ensure fertilization.

How To Grow Muscadines

Generally speaking, Muscadines are trained either on a vertical or an overhead trellis, the vertical trellis constructed much like an arbor for bunch Grapes.

With this system set your posts about fifteen feet by fifteen feet. They should be high—at least seven feet. Wires are then strung across the top of the posts, on the square and on the diagonal. This method will allow space for eight fruiting canes to develop from each Muscadine plant. This will permit more fruiting wood for each vine than would be obtained on a vertical trellis.

Where it is practical to use the overhead system it is very desirable, since it exposes all the plant to full sunlight. Be sure to brace the end posts of the trellis carefully.

Pruning Muscadines

The Muscadine has a boundless enthusiasm for growth, and you must restrain it or you will soon have a jungle of vines. Therefore just as soon as you can, establish a main trunk for the vine. Tie this to the post and cut it off when it reaches the top. The trunk then may be allowed to develop about eight arms near the top. These should radiate outward like the spokes of a wagon wheel. To support them properly, wires should be stretched between the posts, thus forming a canopy. The main arms of the Muscadines do not produce fruiting shoots. One-year-old canes growing from these arms are pruned back to provide fruiting shoots. To prune properly, cut back the previous season's side growth, allowing about six buds to remain on the canes.

For best results, cut out one of the main arms each year. Then select a shoot near the top of the trunk to replace it. If this is done faithfully you will renew all of the arms every eight years. If you don't do this, the old arms in time will become so heavily spurred that their fruiting vigor will be reduced.

CARE OF GRAPE VINES

Good care is always necessary if you would maintain your vines in full vigor, but it is especially important during the formative years when the vines are becoming established. This is when the root systems settle in and the vines build their framework.

My father, who grew up in the Rhineland country of Germany, always was most particular about this. The Germans are famous for being thorough in everything they do, and he neglected nothing toward the end of having his beloved vines produce abundantly.

He saw to it also that my education in Grape culture was most complete. On my grandfather's estate in Germany there had been many, many vines, and Father taught me as his father had taught him. He was very advanced in all his ideas and he required me to learn as much as my four older brothers. Being a curious and intensely-interested little girl, I loved every minute of the time I spent with him tending the vines. And many years later when I had a small vineyard of my own I remembered all these things.

Watering is very important, especially during the summer when the blazing sun beats down day after day, kissing the Grapes into honey-sweetness but burning the humus out of the ground and baking it to brick-like hardness.

During the first growing season water frequently. When the soil becomes dry, soak it deeply and thoroughly. Do not do this, however, when the vines are in bloom. Avoid it also during the last part of the fruiting season. If water is withheld during late summer it will encourage the vines to toughen up or "ripen," and they will not be likely to be damaged by early winter freezing.

Cultivation is very good for Grapes and they respond to this attention. Start in the spring of each year and hoe the soil frequently, at least every two weeks or so until mid-summer.

Should you mulch Grape vines? Many authorities say not to mulch the first few seasons. Unlike many other small fruits, Grapes will develop deep, far-ranging root systems, and they should be allowed to do so before mulching. If you mulch before this happens, the roots are encouraged to stay near the surface. Deep rooting always should be encouraged.

I do not break this rule, but I bend it a little. Most active growing takes place in the spring, and to conserve moisture and keep the ground from drying out completely, I start mulching about the middle of July, pulling it off again the following spring.

Fertilizing: Plenty of manure (well-composted) and bone meal incorporated into the soil at planting time should be sufficient during the first growing season.

After that watch your vines carefully. When they fail to make good growth (yellowing of the leaves may be a sign of nutrient deficiency) fertilize as previously directed.

I firmly believe that every garden can produce good Grapes, for they are one of the easiest fruits to grow and seldom are subject to

---

How To Treat Frost-Injured Vines

A late frost may severely injure the new growth of Grapes. If this happens, remove all new growth—injured and uninjured parts.

The buds of Grapes are compound, and when the first growth is removed, a secondary bud normally develops and produces a partial crop. A few shoots on frosted vines may be uninjured. If only the injured shoots are removed, the uninjured ones will make fast growth.

Very few secondary buds develop from partially stripped vines, so complete stripping is necessary to force secondary growth. Generally, vines that are frosted enough to need some stripping should be stripped completely.

If injury appears to be mild, it is safer not to strip. Such vines will produce a partial crop without stripping, and stripped vines never produce more than a partial crop.

attacks from insects or diseases. If you live in the North where the growing season is short, you can hasten ripening by planting on the south side of a building and training your vines against it. Particularly if the building is painted white or a light color, the Grapes will receive heat radiation from the reflected sunlight, and your fruit may ripen as much as a week or ten days earlier.

As long as you are careful to place your Grapes in a well-drained location and a soil that is fairly deep, they will grow well on soils that are either heavy or light. Actually a soil of just average fertility is best for Grapes, as a very rich condition may stimulate too much cane growth, resulting in clusters that are poorly formed.

A good fertilizer program also may include the application in the fall of two to three pounds of finely-ground granite rock for each vine. This will supply the potash needed for healthy vines and will increase the fruit yields. In the spring a half pound of organic nitrogen, supplied by cottonseed, soybean meal or blood meal also is good. Spread this evenly over an area six to eight feet from the base of the vine, for a mature Grape vine may have roots that spread as far as eight feet from its base. Feeder roots are usually three to six feet away from the base, so spreading the fertilizer this distance from the trunk will be of more benefit than if it is placed close in.

Something else to consider if you live in the North is the height of your trellis. Grapes grown on a trellis where the top wire is five and one-half feet from the ground and the second about two feet below this, will ripen a week or so earlier than those on a shorter trellis. This is because more leaf area is exposed to sunlight.

INSECTS AND DISEASES

If you will grow Grapes that are adapted to your climate, you have taken the first step toward disease and insect control. Any plant that grows vigorously and well, putting its whole strength into it, will throw off troublesome pests better, and diseases may never bother it.

Pest and disease control includes selection of a suitable planting site, the use of disease-resistant varieties and the purchase of healthy, disease-free planting stock. Buy from reputable nurseries, which are proud of their reputations and do everything possible to maintain a high standard.

Good pruning practices, which do not allow the vines to grow too thickly, will go a long way toward control of diseases such as **black rot**, which often is brought on by humid conditions and poor air circulation.

Good circulation also is important in the control of **mildew** on Grapes, for mildew spores cannot tolerate dry, clear air that is warm and moving naturally. Again the selection of a good site will help to keep mildew from becoming destructive.

The chewing insects, **rose chafers**, sometimes are a problem, and if present in considerable numbers can do a lot of damage. Clean cultivation controls them, since their grubs feed on grass roots. **Japanese beetles** attack Grapes in some areas, and again I suggest Milky Spore Disease. The **Grape leafhopper** can be controlled with rotenone spray.

Nicotine sulfate, an old-time material still available, is an extract treated with sulfuric acid. It is less toxic than free nicotine, and when used, highly diluted with water, a soap solution may be added to make the mixture spread and adhere better. It is useful against small soft-bodied insects such as **aphid**, **whitefly**, **leafhopper**, **psylla**, **thrip** and **spider mite**.

Ryania, a plant-derived insecticide, has been found to be effective against Japanese beetles, several species of aphids and the larvae of Oriental fruit moths. Ryania may be used as a dust or spray.

If insect infestations are not very severe, I often have found that a strong spray of just plain water will prove effective. Occasionally I add soap and sometimes this helps, making unnecessary the use of anything stronger.

Birds and wasps can be a great nuisance when Grapes start to ripen. You can protect them with nylon net—the type that is 72 inches wide and comes in many colors. Green of course is preferable, but any color will do. Sew together two pieces, making a piece 154 inches wide and the strip as long as your arbor. If you lace the two lengths together with sturdy nylon yarn you can take it apart at the end of the season and use the strips separately in the spring over your strawberry beds. This netting will also foil Japanese beetles and any other small neighborhood pilferers, which may come over from a nearby property.

GRAFTING

Sometimes it may be desirable to graft a variety of Grapes presently growing in your vineyard to a more desirable type. Do this in the spring just before the sap starts to flow.

A cleft graft is used for this purpose, usually made during February or March. Pieces of a one-year-old cane of a desirable variety are cut in lengths, each containing two or three buds, and are prepared before the sap flow starts.

These scions should be held in damp sawdust or sand, or buried in the soil on the north side of a building so they will remain moist until time for use.

The lower end of the scion is cut wedge-shape by making the cuts about one and one-half to two inches long at the lower end of the scion.

Remove the soil from around the old grape vine to a depth of two or three inches.

At this place the vine is cut or sawed off and either split or sawed vertically to a depth of about two inches. A screw driver, a wedge-shaped tool or bit of wood cut wedge-shaped is placed in the split holding the vine halves apart until the scions can be inserted between the two sides, bringing the cambium layer (the plant's vital tissues that is just under the bark) of each in contact.

If the scions are prepared slightly wedge-shaped the contact will be better made. The screw driver or bit of wood used to hold the halves apart is then removed and the old vine should have sufficient spring to hold the scion in place.

If the halves appear loose, use a strong piece of twine to bind the stock. A piece of heavy paper then should be wrapped around the stock and scion, and soil mounded over the area to prevent drying out.

The top bud of the scion should be about level with the top of the soil. Remove the sucker sprouts that usually appear when an old vine has been cut off. After growing for a year or two the new top of the grafted variety should start producing grapes.

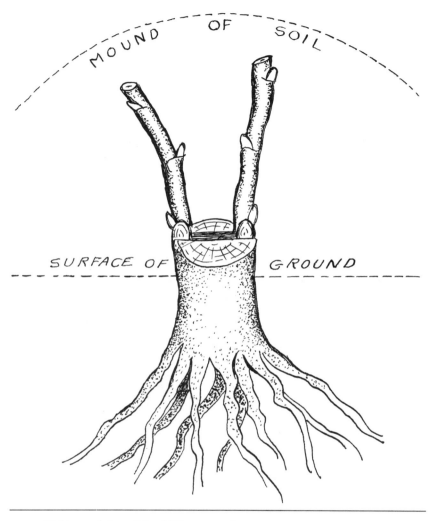

Making a cleft graft in old Grape trunk.

MAKING GRAPE CUTTINGS

Grape cuttings are made by selecting pencil-size canes of the wanted varieties. Cut these canes in lengths containing about three or four buds each. Cuttings are usually made up at the time the grape vines are pruned in January or February. Bundles containing about twenty-five cuttings should be tied together and labeled. Point all buds on the cuttings in the same direction in the bundle.

Bury the bundles with the buds pointing downward (upside down) in the soil in a well-drained area and cover them with about six or eight inches of soil.

In the spring, after the danger of heavy frosts has passed, the cuttings are dug up and set in the nursery rows. Have the soil well prepared. Push the cuttings (right side up this time) down into the soil at about a 45-degree angle, leaving the top bud above the ground layer.

Unless the ground is quite wet, water the cuttings well and then firm the soil about them. Space the cuttings about six inches from each other. Rows may be three or four feet apart depending upon the cultivating equipment you will use. A good dust mulch form of cultivating is followed throughout the growing season.

Cuttings produced in this manner usually will result in well-rooted plants that can be set in the vineyard early the following spring.

Some people have found it advantageous to grow a few Grape cuttings to replace or add to their vineyard, selecting the varieties that are best suited to their climate and conditions.

Therefore, if you have a vine that does particularly well for you, bearing well and being relatively free of insects and diseases, this is a good way to increase your stock while possibly discarding some others that have proven impractical.

RECIPES

Unfermented Grape Juice
Use sound, well-ripened Grapes. Wash in cool water, drain and crush. If additional liquid is needed add one-half cup of water to each pound of Grapes.

Heat Grapes and water to 170°F to 180°F, and keep at this temperature until the juice can be separated from the pulp, for about four or five minutes.

Strain mixture through several thicknesses of clean white cheese-cloth. Strain a second time for clearer juice. Add one-half to one cup of sugar for each gallon of strained juice.

Heat juice to 170°F and fill hot sterilized quart jars to within one-eighth inch of top. Seal and process 30 minutes in hot water bath at simmer. Remove and complete seal.

Grape Jelly

Wash Grapes, stem and crush. Add a small amount of water (barely enough to cover) and boil 15 minutes. Press through a jelly bag (or a fruit press) and strain. To prevent crystals from forming in Grape jelly, allow juice to stand overnight in a cool place or in refrigerator. Next morning carefully pour off juice, leaving sediment in bottom of container.

Grape jelly may be made by following the recipe given with one of the commercial pectins, or:

Measure juice and bring to a boil. For each cup of juice add ¾   cup of sugar. Boil rapidly to jelly stage. Remove at once from the heat.   Pour into dry, sterilized glasses and seal with paraffin.

Crystallized Grapes

This is one of the prettiest possible ways to serve Grapes for dessert or they may be used as a decoration for the center of the table.

Carefully wash a pound of red or blue Grapes and dry on paper towels. Cut the large clusters into several smaller ones. Boil a half cup of water and one cup of sugar for five minutes.

Dip each bunch of Grapes into slightly cooled syrup, holding over pan a minute or two so that excess syrup will drain. Sprinkle Grapes at once with coarse granulated sugar. Place clusters on tray an inch or so apart and allow syrup to harden. If rapid crystallization is desired, chill the Grapes.

Grape Tarts

½ cup Grape juice
1 cup sugar
1 teaspoon cornstarch
1 cup Grapes (any variety)
6 tart shells (baked)
1 cup whipped cream (or whipped Milnot)
½ cup water

Boil together Grape juice, sugar and ½ cup water. Using a little cold water, blend cornstarch to a smooth paste and stir in. Cut Grapes in halves and, if seeded variety, remove seeds. Put in syrup and reduce heat. Simmer until Grapes are soft. Turn into tart shells and top with whipped cream. Top each tart with one uncooked Grape.

Note: Add 1 teaspoon of dissolved gelatin to ½ pint of whipping cream. Chill and use a rotary beater for whipping. Cream will remain firm much longer.

Grape Puree

Wash and de-stem Grapes. Heat Grapes 8–10 minutes at low temperature (not over 180°F), to loosen skins. Do not boil. Put through food mill or wide-meshed strainer. Discard skins and seeds. Grape puree may be canned or frozen. For canning, process as you would Grape juice.

Grape Table Syrup

1 ½ cups Grape puree or Grape juice
1 ½ cups sugar
¼ cup corn syrup
1 tablespoon lemon juice

Combine ingredients. Bring to a rolling boil. Boil one minute (count time after mixture comes to a boil that cannot be stirred down).

Remove from heat; skim off foam. Pour into pre-heated half-pint jars. Cool; cover and store in refrigerator or process in boiling water-bath canner for 10 minutes. Yield: 2 half-pints. Great with waffles or pancakes.

Grape Cheese Cake

2 cups Grape puree
2 packages gelatin (unflavored)
1 cup sugar
3 cups creamed cottage cheese
red food coloring (optional)
1 ½ cups graham cracker crumbs
6 tablespoons melted butter or margarine
¼ cup sugar

Soften gelatin in ½ cup of Grape puree. Heat remaining puree and add gelatin mixture and sugar. Stir until dissolved. Allow mixture to cool slightly. Beat cottage cheese until creamy, using electric mixer at high speed. Add gelatin mixture and several drops of food coloring. Mix until all ingredients are of creamy consistency. If electric blender is used, place cottage cheese, gelatin mixture and red coloring in blender. Blend at high speed until mixture is creamy. Cool until mixture begins to thicken and pour into graham cracker crust.

Graham Cracker Crust

Combine graham cracker crumbs, sugar and melted margarine. Save ½ cup of crumb mixture. Place remaining mixture in a deep 9-inch round pan. Pat firmly with the palm of your hand against bottom and sides of pan to form a shell. Bake crust in 375°F oven for 15 minutes. Chill crust and fill with cheese Grape mixture. Refrigerate until firm. Sprinkle remaining crumbs over the top.

Grape Aspic

2 tablespoons unflavored
   gelatin
½ cup orange juice
¼ cup lemon juice
½ cup sugar

½ teaspoon salt
2 cups Grape puree
1 cup water
2 whole sticks cinnamon
2 whole cloves
red food coloring (optional)

Sprinkle gelatin over mixture of orange and lemon juice to soften. Combine sugar, salt, Grape puree, water and spices (tied loosely in a cheesecloth bag and lightly pounded). Heat 15 minutes; do not boil. Remove spice bag, add gelatin mixture and stir to dissolve. Add a few drops of red food coloring. Refrigerate until firmly set.

Grape Ketchup

2 quarts fresh Grapes
2 cups sugar
1 cup vinegar
½ teaspoon salt
1 stick cinnamon
1 tsp. whole cloves
1 tsp. whole allspice

Crush and simmer Grapes in their own juice until tender. Press through sieve or food mill (discard seeds and skins). Add vinegar, salt and spices (tied loosely in cheesecloth bag and lightly pounded). Cook rapidly until thick. Stir to prevent sticking. Remove spice bag. Pack into pre-heated pint jars, leaving ½ inch head space. Wipe jar mouths and adjust lids. Process in boiling water-bath canner 5 minutes. Yield: about 3 pints.

Rose hips

10.

Berry Novelties for Fun and Fruit

HIGH BUSH CRANBERRY

he High Bush Cranberry is not a true Cranberry, but variety *Viburnum Opulus,* the European Cranberry, is a rugged shrub which will do well in sun or partial shade. It grows six to eight feet tall and has dark green leaves somewhat resembling the maple's.

The bright berries, a flaming scarlet when fully ripe, are used to make a tasty jelly. High Bush Cranberry, whose vitamin C content is very high, is both useful and ornamental, and it is one of the few berries seldom bothered by birds.

Little pruning is needed, some cutting out of the oldest canes close to the ground being all that is necessary. Pruning, which should be done in winter, encourages the growth of renewal shoots from the base of the plant.

BUFFALO BERRY

Buffalo Berry (*Shepherdia*) (Gurney's), a spiny shrub with silvery leaves, bears great clusters of edible red berries that make delicious jelly. This shrub, often used as a windbreak, is extremely hardy and may be grown far north in the

United States and up into Canada. It has the further virtue of with-standing poor soil and dry, windswept locations as well as extreme cold.

The species is dioecious—that is, male and female plants must be included in a planting or the bushes will not bear. Its propagation is by seeds, suckers or hardwood cuttings taken in the fall and they are handled much the same way as are grape cuttings.

ELDERBERRIES

The cultivated Elderberry, Adams (Kelly Bros.), is vastly superior to common kinds. It has great, luscious clusters of glistening black fruit which are great for pies or for combining with other fruit. And does anyone remember the Elderberry wine that Grandmother used to make? My grandmother always had a few reserve bottles on hand to take to sick friends. Elderberries are gregarious. Two should always be planted for cross-pollination. The plant is inclined to ruggedness and may grow as tall as twelve feet in height, flowering in late June and ripening its dark red, nearly black berries in late August. Propagation is by hard wood cuttings.

The wild ones like to grow in moist places, so take a hint from that if you plant it on the home grounds. But though it likes moisture, it must have good drainage. The plants will not do well where the soil is waterlogged. Prune to thin out old wood, to prevent crowding.

Spiced Elderberries

1 stick cinnamon
1 tablespoon whole cloves
1 tablespoon whole allspice
3 pounds sugar
1 pint diluted vinegar
5 pounds Elderberries

Tie spices in cheesecloth bag. Heat sugar, vinegar and spices to boiling and cool. Add clean berries, heat slowly to simmering and simmer until berries are tender. Cool at once and let stand several hours or overnight. Remove spice bag. Pack berries into sterilized jars, Heat syrup to boiling and pour over berries. Seal at once, using pint jars.

HUCKLEBERRIES

The name "Huckleberry" is sometimes applied to Blueberries, but this is in error—as any New Englander will tell you. Blueberries have numerous small seeds, as many as 60 or 70, which are so tiny as to be barely noticeable. Huckleberries have ten comparatively large seeds which you certainly *can* notice, for they will crackle between your teeth. They also have small yellowish dots on the undersides of their leaves which Blueberries do not have.

Huckleberries are borne in clusters, and when fully ripe are a shining black color. In size they are about one-half to three-quarter inch in diameter. Fruits are not edible until fully ripe, at which time they are a jet black and thoroughly soft, as they would be after a first frost.

Huckleberries are not good for eating fresh but are great for making pies and preserves. They also are excellent for freezing and canning.

Garden Huckleberry (Burgess) will grow and thrive in almost any climate. Culture is much the same as for tomatoes. They will do best in a soil that tends toward acidity, and may be propagated by seeds sown in sandy peat, or by cuttings made in summer and inserted in similar soil in a closed propagating frame.

Huckleberry Pie

4 cups fresh Huckleberries
2 tablespoons granulated tapioca
1 cup sugar (½ white, ½ brown)
⅛ teaspoon salt
1 tablespoon lemon juice
1 recipe Plain Pastry
1 tablespoon butter or margarine

Stem and wash berries. Mix tapioca, sugar and salt together and sprinkle over berries; add lemon juice. Line pie pan with pastry, pour in filling and dot with margarine. Cover with top crust. Bake in very hot oven, 450°F for 10 minutes, reduce temperature to moderate, 350°F, and bake 35 minutes longer, or until berries are tender.

TAYBERRY

This new hybrid (Stark Bros.) from the Highlands in Scotland is gaining favor widely. Scottish researchers crossed a Blackberry with a red Raspberry, resulting in a fruit with a delightful flavor quite different from either parent. Tayberries are very fruity and sweet. They make large plants and should be planted 6 feet apart in rows. Care for them as you would Blackberries. Tayberries are very productive: you can pick the 1-inch dark red berries by the bucketfuls starting in early July in zone 6. In zones 5–8 where temperatures go below 15 F they will need protection. They are only shipped in the spring.

SERVICEBERRY

The Shadblow Serviceberry (Henry Field), sometimes also called Juneberry or May Cherry, is a beautiful shrub which bears small white flowers. It grows six to eight feet tall and is a mighty bearer of delicious "blueberry-size" berries. These were much used by various American Indian tribes alone and in pemmican, and we use them now for jams, sauces and pies.

The fruits are reddish-purple to almost black when fully ripe and often are produced the first year after planting.

Plants of Serviceberries may be purchased or they may be propagated by sowing seeds outdoors in October. Some types may be increased by lifting and dividing the clumps in autumn after the leaves have fallen.

GROUND CHERRY

This is sometimes called Husk Tomato, Strawberry Tomato or Poha Berry. It is classed as a tomato even though it has an odd, papery husk enclosing each fruit. And it has a taste all its own.

The Ground Cherry occurs naturally from Massachusetts to Florida and westward. It is easily cultivated in a well-drained soil if given a sunny location. In warm climates the seeds may be sown directly out of doors. In colder areas start the seeds indoors three weeks before time to set them in open ground. Cover seeds thinly, pack firmly and water.

When the plants are about two inches high, transplant to three inches apart in a flat or pot. After danger of frost is past, set two to three feet apart in the garden. Before doing this it's a good idea to harden the plants by gradually exposing them to the outdoor air for about a week. The plants have a rather sprawling habit of growth, so if you have more than one row, allow plenty of room between the rows.

Ground Cherries may be used in a number of different ways, one of which is for a delicious preserve.

Ground Cherry Preserves

Remove the husks from the Ground Cherries. Make a syrup of one cup white sugar, and one-half cup brown sugar (or ½ cup honey), three cups water and the juice of two lemons (you may also grate lemon rind and add this).

Boil syrup 5 minutes and then add enough Cherries to come to the top of the syrup. Boil slowly until Cherries are tender and clear. If you have a quantity you can place in sterilized jars and seal as with other fruits.

Ground Cherry Pie

1 ½ quarts cleaned fruit
Juice of 2 lemons
½ teaspoon cinnamon
¼ teaspoon ginger
1 cup white sugar
½ cup brown sugar
½ cup flour (preferably unbleached white)

Break open fruit but do not separate. Place in saucepan with other ingredients and bring to a boil. Pour into unbaked pie shell, dot with margarine, and sprinkle on a pinch of salt.

Add top crust and pierce with a fork. Bake in a 500°F oven for 5 minutes. Reduce heat to 350°F and bake until done, about 40 minutes.

ROSE RUGOSA

In recent years *Rosa Rugosa*, the "hippy" rose, has attracted a great deal of attention largely because of the high content of Vitamin C contained in the large, meaty, rose hips.

The Rugosa blossoms are single-petaled, very lovely, and the bushes, closely planted, make an attractive and nearly impenetrable hedge. The rosy flowers are followed by the hips which turn ruby red when fully ripe, again making the bushes exceedingly attractive.

These bushes may be ordered from several nurseries that specialize in these lovely plants. *Rosa Rugosa Rubra* (Hastings) has magenta flowers with a spicy fragrance. The bright orange hips are borne abundantly. Two or more plants are needed for cross-pollination to produce the most hips.

When your plants arrive, set them out as soon as possible. For individual plants dig a hole of sufficient size to accommodate the roots without crowding, but if you plan a hedge dig a long trench about one foot across and a foot deep.

Place some well-rotted manure or compost several inches below where the roots will rest. This will help your plants to get a quick start and produce sooner. Actually, this Rose is very hardy and needs little care. To help it get established, cut back the stems leaving three or four buds or leaf nodes on each stem retained. The young plant will benefit from mulching especially in hot, dry climates.

Rose Hip Puree

To receive the most benefit from the hips, which are produced only after the blooms and petals have fallen, gather them when they are fully ripe, but not overripe—usually after the first frost. If they are orange, it is too early. If they are dark red, it is too late. Gather them when they are a brilliant scarlet.

When ready to process, trim both ends of the rose hips with a pair of kitchen scissors before cooking.

Use stainless steel knives, wooden spoons, earthenware or china bowls, and glass or enamel saucepans. Do *not* use copper or aluminum utensils.

Cook rapidly with cover on vessel to prevent loss of Vitimin C. Then strain out the spines and seeds, or break them down by rubbing the cooked pulp through a sieve or food mill. The puree obtained may be used for jams or jellies.

Rose hips also may be dried and kept for long periods, to be ground into powder, which may be added to other foods and drinks, to waffles and pancakes, for instance. The powder blends well but adds little flavor. Thus we get the vitamin benefit without having tastes changed.

Many rose hip products now may be obtained from commercial sources. These include wines, soups, jams, rose honey, rose vinegar, rose syrup, rose sugar, dried rose hips, canned rose hips—and even beauty lotions for the face and hands.

Drying Berries

Drying, according to Grace Firth in *A Natural Year*, is the simplest and cheapest manner of preserving berries: "Handled properly, they will keep a superior flavor." Substitute reconstituted dry berries for canned fruit in pies, shortcakes, and sauces.

To dry berries, use firm, nearly ripe but not mushy fruit; do not wash. Put them one layer thick into an oven and thoroughly heat at 150°F for about 20 minutes. Do not hold them in the oven long enough to cook them. When heated enough to destroy undesirable bacteria, spread thinly on enamel, wood, or stone and cover them with netting. Take inside at night to prevent sweating. Ten days of good sunlight should dry most berries, depending on kind or size.

"Store in sterilized, covered jars, which may," she writes, "but need not be vacuum-sealed." Soak overnight in water to reconstitute; or boil gently for 30 minutes, adding a little sugar and a wee sprinkle of salt to bring them to life.

A FINAL WORD

I seem to have come to a stopping place. There it is—the sum total of my own fifty or more years of experience in growing berries, not from the standpoint of a scientist but just as a shirt-sleeve gardener. And I'll bet this book no sooner gets into print than there will be a rush of conflicting opinions—and quite probably we both will be right. This is because there are so many variables—of soil, climate and inclination—for the would-be grower-of-berries to give his time and attention.

But remember this: in nothing do I take an arbitrary stand. I do not feel that my way is the only way. I do think that a knowledge of the basics is important, for this at least gives us something to depart from if contrary opinion develops.

I urge you to do as I have done—to experiment. The location of your own particular homestead or suburban property, because of some physical feature of the land, may create entirely different situations from your neighbor's. It may be higher or lower, wetter or drier. You may be able to grow every fruit mentioned in this book, or you may need to be very selective.

A word, too, about insect pests and diseases. Though these are mentioned for the sake of identification and possible remedies, I do not feel that great emphasis should be placed on them. Don't let them frighten you away from growing small fruits and berries, for most of the diseases will never happen. All reputable nurseries today sell only virus-free plants, and if you destroy old bramble bushes nearby, infection is unlikely. As for insect pests, they are never entirely eradicated, but seldom do their numbers increase to such proportions, largely due to the diversification in the planting of home gardens, that they become a real menace. Natural controls keep them down. The birds, which help in this but love berries too, can be prevented from serious pilfering. Good soil preparation, choice of good stock suited to your climate and reasonably good routine care should add up to success.

Down through the years my garden always has been the source of pure pleasure to me, both in its care and in the enjoyment of its fruits. I wish the same for you.

Appendix

SOURCES OF SMALL FRUIT PLANTS

This is a partial list of firms selling small fruit plants. The inclusion of a firm is not a guarantee of reliability, and an absence does not imply disapproval. These addresses were viable at the time this book was written; the author is not responsible for changes of address or discontinued firms or varieties.

Armstrong Nurseries
P.O. Box 4060
Ontario, CA 91761

Bountiful Ridge Nurseries, Inc.
Princess Anne, MD 21853

Burgess Seed & Plant Co.
Bloomington, IL 61701

W. Atlee Burpee & Co.
Warminster, PA 18974

Farmer Seed & Nursery Co.
Faribault, MN 55021

Henry Field Seed & Nursery Co.
Shenandoah, IA 51602

Gurney Seed & Nursery Co.
Yankton, SD 57079

Earl May Seed & Nursery
Shenandoah, IA 51603

Mellinger's
North Lima, OH 44452-97301

Hastings
P.O. Box 115535
Atlanta, GA 30310-8535

Johnny's Selected Seeds
Foss Hill Road
Albion, MN 04910

Kelly Bros., Nurseries, Inc.
Dansville, NY 14437

Lakeland Nurseries Sales
Hanover, PA 17331

McFayden Seed Co., Ltd.
Brandon, Manitoba
R7A 6NA Canada

Nichols Garden Nursery
1190 North Pacific Highway
Albany, OR 97321

Geo W. Park Seed Co., Inc.
Greenwood, SD 29647-0001

R.H. Shumway
Rockfield, IL 61101

Spring Hill Nurseries
Tipp City, IA 45366

Stark Bros.
Louisiana, MO 63353

Stern's Nurseries
Geneva, NY 14456

Stokes Seeds, Inc.
Buffalo, NY 14240

OF SPECIAL NOTE

The Seed Savers Exchange, dedicated to preserving heirloom plants, has put together an exhaustive guide to nurseries carrying heirloom fruits, berries, and nuts. The *Fruit, Berry, and Nut Inventory*, edited by Kent Whealy is available for $19 by writing Seed Savers Exchange, Rural Route 3, Box 239, Decorah, Iowa 52101.

Although yield, spacing, and date of maturity will necessarily vary within each type of small fruit, this is a good approximation of what may be expected.

Fruit	Planting distance		Interval from planting to fruiting	Life of plants	Height of mature plant	Estimated annual yield per plant	Suggested number of plants for family of 5
	between rows	between plants					
	(feet)	(feet)	(years)	(years)	(feet)		
Strawberries (matted row)	4	2	1	3–4	1	½–1 qt. per ft. of row	100
Currants	6–8	4	2	12–15	3–4	3 quarts	4–6
Gooseberries	6–8	4	2	12–15	3–4	4–5 quarts	4–6
Raspberries—red	6–8	3–4	1	8–10	4–5	1½ quarts	20–25
—black	6–8	3–4	1	8–10	4–5	1 quart	20–25
—purple	6–8	3–4	1	8–10	4–5	1 quart	20–25
Blackberries—erect	6–8	4–5	1	10–12	3–5	1 quart	15–20
trailing, or semi-trailing, includes Boysenberries, Dewberries, etc.	(Staked or trellis) 6–8	6–10	1	8–10	6–8	4–10 quarts	8–10
Blueberries	8–10	6–8	2	20	6–10	3–4 quarts	8–10
Grapes	8–10	8–10	3	20	6	¼–½ bushel (trellised)	5–10
Everbearing strawberries (hills)	1–1½	1–1½	½	2–3	1	½ quart	100
Everbearing raspberries	8	3	½	8–10	4–5	1 quart—spring ½ quart—fall	15–20

These yields, of course, are representative of plants which have had good care and are grown out well.

Nutritive Value of Small Fruits

Fruits (1 cup)	Grams	Percent Water	Calories Food Energy	Grams Protein	Grams Fat (total lipid)	Grams Carbohydrate	Milli-grams Calcium	Milli-grams Iron	Units Vitamin A Value	Milli-grams Thiamine	Milli-grams Riboflavin	Milli-grams Niacin	Milli-grams Ascorbic acid
Blackberries, raw	144	84	85	2	1	19	46	1.3	290	.05	.06	.5	30
Blueberries, raw	140	83	85	1	1	21	21	1.4	140	.04	.08	.6	20
Grapes, raw; American type (slip skin) such as Concord, Delaware, Niagara, Catawba, and Scuppernong	153	82	65	1	1	15	15	.4	100	.05	.03	.2	3
European type (adherent skin), such as Malaga, Muscat, Thompson Seedless, Emperor and Flame Tokay	160	81	95	1	Trace	25	17	.6	140	.07	.04	.4	6
Raspberries, raw	123	84	70	1	1	17	27	1.1	160	.04	.11	1.1	31
Strawberries, raw, capped	149	90	55	1	1	13	31	1.5	90	.04	.10	1.0	88

Index